Nonsense on Stilts

Nonsense on Stilts

How to Tell Science from Bunk

MASSIMO PIGLIUCCI

THE UNIVERSITY OF CHICAGO PRESS CHICAGO AND LONDON

MASSIMO PIGLIUCCI is professor of philosophy at the City University of New York—Lehman College. He is the author of several books, including, with Jonathan Kaplan, *Making Sense of Evolution*.

The University of Chicago Press, Chicago 60637
The University of Chicago Press, Ltd., London

Printed in the United States of America
19 18 17 16 15 14 13 12 11 10   3 4 5

ISBN-13: 978-0-226-66785-0 (cloth)
ISBN-10: 0-226-66785-5 (cloth)
ISBN-13: 978-0-226-66786-7 (paper)
ISBN-10: 0-226-66786-3 (paper)

Library of Congress Cataloging-in-Publication Data

Pigliucci, Massimo, 1964–
Nonsense on stilts : how to tell science from bunk / Massimo Pigliucci.
p. cm.
Includes bibliographical references and index.
ISBN-13: 978-0-226-66785-0 (cloth : alk. paper)
ISBN-13: 978-0-226-66786-7 (pbk. : alk. paper)
ISBN-10: 0-226-66785-5 (cloth : alk. paper)
ISBN-10: 0-226-66786-3 (pbk. : alk. paper) 1. Pseudoscience—History. 2. Science—History. I. Title.
Q172.5 .P77P54 2010
500—dc22

2009049778

⊗ The paper used in this publication meets the minimum requirements of the American National Standard for Information Sciences—Permanence of Paper for Printed Library Materials, ANSI Z39.48-1992.

Contents

INTRODUCTION

Science versus Pseudoscience and the "Demarcation Problem"

"The foundation of morality is to . . . give up pretending to believe that for which there is no evidence, and repeating unintelligible propositions about this beyond the possibilities of knowledge." So wrote Thomas Henry Huxley, who thought—in the tradition of writers and philosophers like David Hume and Thomas Paine—that we have a moral duty to distinguish sense from nonsense. That is also why I wrote this book. Accepting pseudoscientific untruths, or conversely rejecting scientific truths, has consequences for all of us, psychological, financial, and in terms of quality of life. Indeed, as we shall see, pseudoscience can literally kill people.

Yet, the unstated assumption behind Huxley's position is that we can actually tell the difference between sense and nonsense, and in the specific case that concerns us here, between good science and pseudoscience. As it turns out, this is not at all an easy task, and it requires an understanding of the nature and limits of science, of logical fallacies, of the psychology of belief, and even of politics and sociology. It is this search for understanding that constitutes the bulk of what you are about to read, but the journey should certainly not end with the last page of this book. Rather, I hope the reader will use the chapters that follow as a springboard toward even more readings and discussions, to form a habit of always questioning with an open mind and of constantly demanding evidence for whatever assertion is being made by whatever self-professed authority—including of course yours truly.

The starting point for our quest is what mid-twentieth-century philosopher of science Karl Popper famously called "the demarcation problem." Popper wanted to know what distinguishes science from nonscience, including, but not limited to, pseudoscience. He wanted to arrive at the essence of what it means to do science by contrasting it with activities that do not belong there. Popper rightly believed that the public—not just scientists or philosophers—needs to understand and appreciate that distinction, because science is too powerful and important, and pseudoscience too common and damaging, for an open society to afford ignorance on the matter.

Like Popper, I believe some insight into the problem might be gleaned by considering the differences between fields that are clearly scientific and those that just as clearly are not. For instance, we all know of perfectly good examples of science, say physics or chemistry, and Popper identified some exemplary instances of pseudoscience to compare them with. Two of those he considered in some detail are Marxist theories of history and Freudian psychoanalysis. The former is based on the idea that a major key to understanding history is the ongoing economic struggle between classes, while the latter is built around the claim that unconscious sex drives generate psychological syndromes in adult human beings.

The problem is, according to Popper, that those two ideas rather paradoxically explain things a little too well. Try as you may, there is no historical event that cannot be recast as the result of economic class struggle, and there is no psychological observation that cannot be interpreted as being caused by an unconscious obsession with sex. The two theories, in other words, are too broad, too flexible with respect to observations, to actually telling us anything interesting. If a theory purports to explain everything, then it is likely not explaining much at all. Popper claimed that theories like Freudianism and Marxism were unscientific because they were "unfalsifiable." Falsifiability here is the property that a theory has of being capable of being proven wrong by some conceivable observation or experiment. For instance, if I say that adult dogs are quadrupedal (four-legged) animals, my statement is potentially falsifiable: one can easily go around observing dogs and taking notes on whether they are quadrupeds or bipeds (two-legged), and the observations would settle the matter empirically. Either my statement is true or it isn't, but it is a scientific statement—according to Popper—because it is potentially falsifiable by observations.

Rather counterintuitively, Popper also thought that scientific theories cannot ever be proven, because they are always open to the possibility that a new observation—hitherto unknown—will falsify them. For instance, I could observe thousands of four-legged dogs and grow increasingly confident that my theory is right. But then I could turn a corner and see an adult two-legged dog: there goes the theory, falsified by one negative result, regardless of how many positive confirmations I had on my notepad up to that point. In this view of the difference between science and pseudoscience, then, science makes progress not by proving its theories right—because that's impossible—but by eliminating an increasing number of wrong theories. Pseudoscience, however, does not make progress because its "theories" are so flexible that they can accommodate any observation whatsoever, which means that pseudoscientific theories do not actually have any explanatory teeth.

Most philosophers today would still agree with Popper that we can learn about what science is by contrasting it with pseudoscience (and nonscience), but his notion that falsificationism is *the* definitive way to separate the two has been severely criticized. A moment of reflection reveals that Popper's solution is a bit too neat to work in the messiness of the real world. Consider again my example of the quadrupedal dogs, for instance. Suppose we do find a dog with only two legs, perhaps because of a congenital defect. Maybe our bipedal dog has even learned to move around by hopping on its two hind legs, like kangaroos do (in fact, such abnormal behavior has occasionally been observed in dogs and other normally quadrupedal mammals; just google "bipedal dog" to see some amusing examples of it). According to Popper we would have to reject the original statement that dogs are quadrupeds as false. But as any biologist, and even simple common sense, would tell you, we would be foolish to do so. Dogs *are* quadrupeds, exceptions notwithstanding. However, because of the new observation we need to slightly modify our original position and say that dogs are normally quadrupedal animals, adding that occasionally one can find specimens with developmental abnormalities that make them look and behave like a biped.

Similarly, philosophers agree that scientists do not (and should not) in fact reject theories just because they may initially fail to account for some observation. Rather, they keep working at them to see why the data do not fit and what can be done to modify the theory in a sensible manner. Perhaps the new observations do not really challenge the core of a theory because they are exceptional, or maybe the discrepancies arise from the fact that the often sophisticated instruments used by scientists (think of giant telescopes or atom-smashing machines) may be giving incorrect readings and need to be fixed, and so on. This idea of modifying our statements according to new data sounds reasonable on the face of it, until you realize that we are apparently about to slip back precisely into the sort of nonscience that Popper was worried about: if we can tweak our theories at will as new observations come in, we end up with something not really that different from Freudianism or Marxism.

So, on the one hand a strict criterion of falsificationism seems to throw away the proverbial baby with the bathwater: too much good science would fail the falsification test. On the other hand, if we allow ourselves the luxury to change theories to accommodate whatever new observation comes our way we are not doing science at all, but rather engaging in a sterile exercise of rationalization. What are we to do? That is one of the things that we are about to find out in this book: we will search for a more sensible conception of science that will still allow us to tell the difference between it and bunk. In the process, we will see that not all sciences are cast from the same mold: some rely heavily on experiments, while others are more historical in nature and require the attitude and methods of a crime detective to be successfully practiced. We will examine areas of inquiry that are at the borderlands between science and nonscience, then jump across that fuzzy boundary and look at claims that the philosopher Jeremy Bentham would have famously labeled "nonsense on stilts," meaning a really, really tall order of nonsense.

Our voyage will lead us to discussions of how science is presented and misrepresented in the media, in politics, and even in courtrooms—three arenas where science and society interact with profound effects on each other. We will briefly look at the tortuous intellectual history of science since its dim beginnings in ancient Greece; examine issues surrounding the modern practice of science and the widespread acceptance of pseudoscience, such as what it means to be a scientist and intellectual in contemporary society; what science critics have to say about the limits of science and why they matter; and even question the very idea of "experts," in science or otherwise. It is an exploration that will expose us to basic ideas in philosophy of science, as well as to the tools of critical thinking, tools that will be useful in a variety of instances, safeguarding us against being duped by scientists, pseudoscientists, journalists, or politicians. We will see that in the end there is no simple answer to Popper's demarcation problem, but that there are answers nonetheless. Given the power and influence that science increasingly has in our daily lives, it is important that we as citizens of an open and democratic society learn to separate good science from bunk.

This is not just a matter of intellectual curiosity, as it affects where large portions of our tax money go, and in some cases even whether people's lives are lost as a result of nonsense.

Before the fun begins, however, I need to thank many people who have helped and acknowledge some who inspired me, without whom this book would have taken a lot longer to write and would likely not have been as good as I hope it is. Josh Banta, Oliver Bossdorf, Michael De Dora, Gillian Dunn, Mark Jonas, Phil Pollack, and Raphael Scholl have all graciously commented on drafts of several chapters, indubitably increasing their clarity and readability. My editor, Christie Henry, has been enthusiastic and highly supportive of the project from the beginning. The writing of this book has been inspired by a number of people, either through direct discussions or because their writings have been influential on my thinking. In particular, *Skeptical Inquirer* editor Ken Frazier has stirred me to think like a positive skeptic; my philosopher colleague Jonathan Kaplan has enormously augmented my understanding of issues in philosophy of science; and it was reading astronomer Carl Sagan when I was a young student that instilled in me a passion both for science and for writing about it. And while we are talking about people I never met (and never will), the eighteenth-century philosopher David Hume is one of my role models, not just for the clarity and audacity of his thinking, but also because he appreciated the good things in life well enough to deserve the nickname of *le Bon David* from fellow French members of the Enlightenment. Countless, and often nameless, readers of my blog (rationallyspeaking.org) have challenged both my ideas and the form in which I express them, which has been very useful indeed in writing this book. Finally, my wife Gillian has been inordinately patient with me for the two years that it has taken to write this book; I don't know how she does it, but I am very thankful nonetheless.

CHAPTER ONE

Hard Science, Soft Science

You know my methods. Apply them.—Sherlock Holmes, *The Sign of the Four*

You can observe a lot by just watching.—Yogi Berra

"Scientists these days tend to keep up a polite fiction that all science is equal. Except for the work of the misguided opponent whose arguments we happen to be refuting at the time, we speak as though every scientist's field and methods of study are as good as every other scientist's, and perhaps a little better. This keeps us all cordial when it comes to recommending each other for government grants." Candid words about the nature of the scientific enterprise as seen from the inside by a participating scientist. And what makes these sentences even more remarkable is that they were not uttered behind closed doors in a room full of smoke, but printed in one of the premiere scientific magazines in the world, *Science*.[1] It was 1964, the year I was born, and the author was John R. Platt, a biophysicist at the University of Chicago. The debate between scientists on what constitutes "hard" (often equated with good, sound) and "soft" (implicitly considered less good) science has not subsided since, and it provides us with our first glimpse into how difficult—and contentious!—it is to characterize science itself.

Platt was frustrated by the fact that some fields of science make clear and rapid progress, while others keep mucking around without seemingly being able to accomplish much of relevance. As Platt put it, in the same article: "We speak piously of . . . making small studies that will add another brick to the temple of science. Most such bricks just lie around the brickyard." Physics, chemistry, and molecular biology are considered by Platt (and many others) as hard sciences, the quintessential model of what science ought to be. Ecology, evolutionary biology, and other fields like psychology and sociology are soft sciences, and the highest aspiration of people working in these fields is assumed to be to make their disciplines as hard as physics. Platt's article is a classic that should be read by anyone interested in the nature of science, and he was right in pointing out the problem; he was not quite as right in diagnosing its roots, however, and even less so at suggesting a possible cure. Nonetheless, Platt's critique of what others often refer to as soft science provides us with an excellent starting point to explore the idea that, in fact, there may be more than one kind of science, that "science" is a heterogeneous category—a notion that would surprise most in the general public and that will likely be resisted even by a number of scientists. Moreover, our discussion will in turn open up the possibility that there may be ways to identify criteria that not only divide soft and hard sciences, but also separate disciplines that are not quite science yet (and perhaps will never become it) and others that are downright pseudoscientific. This is the task we will pursue over the first three chapters of this book in our quest to explore the complex intellectual landscape first identified by Popper's original demarcation problem.

Strong Inference and the Proper Way to Do Science (or Is It?)

Platt's attack on soft science begins, as we have seen, by stressing the fact that some disciplines seem to make fast and impressive progress, while others have a tendency of going around in circles, or at best move slowly and uncertainly. Before we examine why this is and what could possibly be done about it, a more fundamental question is whether Platt is correct at all in thinking that there is a problem to begin with. It seems clear from even a cursory examination of the history of science that Platt is at least partially correct: some sciences do progress significantly more than others. However, the pattern appears more complex than a simple line dividing "hard" from "soft" disciplines: it is true that, say, particle physics and molecular biology have made spectacular advances during the twentieth century; but it is also true that physics itself went through long periods of stasis on certain problems, for instance the long interval between Newton and Einstein on the question of the nature of gravity. And such periods of slow progress may occur again in the future, even for the "queen" of sciences: for all the talk about a "unified theory of everything," physicists have been trying to reconcile the known discrepancies between their two most successful theories, general relativity and quantum mechanics, for close to a century; they have not succeed yet.

Organismal biology (ecology and evolutionary biology) is often considered a quasi-soft science, and yet it has seen periods of great progress—most obviously with Darwin during the second half of the nineteenth century, and more recently during the 1930s and '40s. Moreover, there is currently quite a bit of excited activity in both empirical and theoretical evolutionary biology, which may be leading to another major leap forward in our understanding of how organisms evolve and adapt to their environments.[2] Molecular biology, on the other hand, hailed by Platt as a very successful hard science on the model of chemistry and physics, may be in the process of running into the limits of what it can achieve without falling back on "softer" and more messy approaches to its subject matter: it is true that the discovery of the structure of DNA in 1953 is one of the all-time landmarks of science; but it is equally clear that the much-touted sequencing of the whole human genome has provided very few hard answers for biologists, instead leading to a large number of "bricks laying around the brickyard," as Platt would have put it. We know a lot more about the human (and other) genomes, but much of what we know is a complex mess of details that is difficult to extricate to achieve a clear picture of how genomes work and evolve.

All in all, it seems that one can indeed make an argument that different scientific disciplines proceed at dramatically different paces, but it is also true that a given science may undergo fits and starts, sometimes enjoying periods of steady and fast progress, at other times being bogged down into a spell of going nowhere, either empirically (lack of new discoveries) or theoretically (lack of new insights).

If we agree that the nature of science is along the lines that I have just described, next we need to ask why this is so. Platt briefly mentions a number of possibilities, which he dismisses without discussion, but that we need to pay some attention to before we move to his main point. These alternative hypotheses for why a given science may behave "softly" include "the tractability of the subject, or the quality of education of the men [*sic*] drawn into it, or the size of research contracts."[3] In other words, particle physics, say, may be more successful than ecology because it is easier (more tractable), or because ecologists tend to be dumber than physicists, or because physicists get a lot more money for their research than ecologists do.

The second scenario is rather offensive (to the ecologists at least), but more importantly there are no data at all to back it up. And it is difficult to see how one could possibly measure the alleged different degree of "education" of people attracted to different scientific disciplines. Nearly all professional scientists nowadays hold a Ph.D. in their discipline, as well as years of postdoctoral experience at conducting research and publishing papers. It is hard to imagine a reliable quantitative measure of the relative difficulty of their respective academic curricula, and it is next to preposterous to argue that scientists attracted to certain disciplines are smarter than those who find a different area of research more appealing. It would be like attempting to explain the discrepancy between the dynamism of twentieth-century jazz music and the relative stillness of symphonic ("classical") music[4] by arguing that jazz musicians are better educated or more talented than classically trained ones. It's a no starter.

The other factors identified and readily dismissed by Platt, though, may actually carry significant weight. The obvious one is money: there is no question that, at least since World War II, physics has enjoyed by far the lion's share of public funding devoted to scientific research, a trend that has seen some setback in recent years (perhaps, not surprisingly, after the end of the cold war). It would be foolish to underestimate the difference that money makes in science (or anywhere else, for that matter): more funds don't mean simply that physicists can build and maintain ever larger instruments for their research (think of giant telescopes in astronomy or particle accelerators in subnuclear physics), but perhaps equally important that they can attract better paid graduate students and postdoctoral associates, the lifeblood of academic research and scholarship. Then again, of course, money isn't everything: our society has poured huge amounts of cash, for instance, into finding a cure for cancer (the so-called war on cancer), and although we have made much progress, we are not even close to having eliminated that scourge—if it is at all possible.

Part of the differential ability of scientific disciplines to recruit young talent also deals with an imponderable that Platt did not even consider: the "coolness factor." While being interested in science will hardly make you popular in high school or even in college, among science nerds it is well understood (if little substantiated by the facts) that doing physics, and in particular particle physics, is much cooler than doing geology, ecology, or, barely mentionable, any of the social sciences—the latter a term that some in academia still consider an oxymoron. The coolness factor probably derives from a variety of causes, not the least of which is the very fact just mentioned that there is more money in physics than in other fields of study, and even the large social impact of a few iconic figures, like Einstein (when was the last time you heard someone being praised for being "a Darwin"?).

Another reason mentioned but left unexamined by Platt is the relative complexity of the subject matters of different scientific disciplines. It seems to me trivially true that particle physics does in fact deal with the simplest objects in the entire universe: atoms and their constituents. At the opposite extreme, biology takes on the most complex things known to humanity: organisms made of billions of cells, and ecosystems whose properties are affected by tens of thousands of variables. In the middle we have a range of sciences dealing with the relatively simple (chemistry) or the slightly more complex (astronomy, geology), roughly on a continuum that parallels the popular perception of the divide between hard and soft disciplines. That is, a reasonable argument can in fact be made that, so to speak, physicists have been successful because they had it easy. This is of course by no means an attempt to downplay the spectacular progress of physics or chemistry, just to put it into a more reasonable perspective: if you are studying simple phenomena, are given loads of money to do it, and are able to attract the brightest minds because they think that what you do is really cool, it would be astounding if you had *not* made dazzling progress!

Perhaps the most convincing piece of evidence in favor of a relationship between simplicity of the subject matter and success rate is provided by molecular biology, and in particular by its recent transition from a chemistry-like discipline to a more obviously biological one. Platt wrote his piece in 1964, merely eleven years after James Watson, Francis Crick, and Rosalind Franklin discovered the double-helix structure of DNA. Other discoveries followed at a breathtaking pace, including the demonstration of how, from a chemical perspective, DNA replicates itself; the unraveling of the genetic code; the elucidation of many aspects of the intricate molecular machinery of the cell; and so on. But by the 1990s molecular biology began to move into the new phase of genomics,[5] where high throughput instruments started churning a bewildering amount of data that had to be treated by statistical methods (one of the hallmarks of "soft" science). While early calls for the funding of the human genome project, for instance, made wildly optimistic claims about scientists soon being able to understand how to create a human being, cure cancer, and so on, we are in fact almost comically far from achieving those goals. The realization is beginning to dawn even on molecular biologists that the golden era of fast and sure progress may be over and that we are now faced with unwieldy mountains of details about the biochemistry and physiology of living organisms that are very difficult to make sense of. In other words, we are witnessing the transformation of a hard science into a soft one!

Despite all of the reservations that I detailed above, let us proceed to tackle Platt's main point: that the difference between hard and soft science is a matter of method, in particular what he refers to as "strong inference." "Inference" is a general term for whenever we arrive at a (tentative) conclusion based on the available evidence concerning a particular problem or subject matter. If we are investigating a crime, for instance, we may infer who committed the murder from an analysis of fingerprints, weapon, motives, circumstances, etc. An inference can be weaker or stronger depending on how much evidence points to a particular conclusion rather than to another one, and also on the number of possible alternative solutions (if there are too many competing hypotheses the evidence may simply not be sufficient to discriminate among them, a situation that philosophers call the underdetermination of theories by the data). The term "strong inference" was used by Platt to indicate the following procedure:

1. Formulate a series of alternative hypotheses;
2. Set up a series of "crucial" experiments to test these hypotheses; ideally, each experiment should be able to rule out a particular hypothesis, if the hypothesis is in fact false;
3. Carry out the experiments in as clear-cut a manner as possible (to reduce ambiguities of interpretation of the results);
4. Eliminate the hypotheses that failed step (3) and go back to step (1) until you are left with the winner.

Or, as Sherlock Holmes famously put it in *The Sign of Four*, "when you have eliminated the impossible, whatever remains, however improbable, must be the truth." Sounds simple enough. Why is it, then, that physicists can do it but ecologists or psychologists can't get such a simple procedure right?

The appeal of strong inference is that it is an extremely logical way of doing things: Platt envisions a logical decision tree, similar to those implemented in many computer programs, where each experiment tells us that one branch of the tree (one hypothesis) is to be discarded until we arrive at the correct solution. For Platt, hard science works because its practitioners are well versed in strong inference, always busy pruning their logical trees; conversely, for some perverse reason scientists in the soft sciences stubbornly refuse to engage in such a successful practice and as a consequence waste their careers scattering bricks of knowledge in their courtyards rather than building fantastical cathedrals of thought. There seems to be something obviously flawed with this picture: it is difficult to imagine that professionally trained scientists would not realize that they are going about their business in an entirely wrong fashion, and moreover that the solution is so simple that a high school student could easily understand and implement it. What is going on?

We can get a clue to the answer by examining Platt's own examples of successful application of strong inference. For instance, from molecular biology, he mentions the discovery of the double-helix structure of DNA, the hereditary material. Watson, Crick, Franklin, and other people working on the problem (such as twice–Nobel laureate Linus Pauling, who actually came very close to beating the Watson-Crick team to the finishing line) were faced by a limited number of clear-cut alternatives: either DNA was made of two strands (as Watson and Crick thought, and as turned out to be the case) or three (as Pauling erroneously concluded). Even with such a simple choice, there really wasn't any "crucial experiment" that settled the matter; instead, Watson and Crick had sufficient quantitative information from a variety of sources (chiefly Franklin's crystallographic analyses) to eventually determine that the two-helix model was the winner. Another example from Platt's article comes from high-energy physics and deals with the question of whether fundamental particles always conserve a particular quantity called "parity." The answer is either yes or no, with no other possibilities, and a small number of experiments rapidly arrived at the solution: parity is not always conserved. Period. What these cases of success in the hard sciences have in common is that they really do lend themselves to a straightforward logical analysis: there is a limited number of options, and they are mutually exclusive. Just like logical trees work very well in classic Aristotelian logic (where the only values that can be attached to a proposition are true or false), so strong inference works well with a certain type of scientific question.

Yet, any logician (and most people endowed with common sense) knows very well that the realm of application of Aristotelian logic is rather limited, because many interesting questions do not admit of simple yes/no answers. Accordingly, modern logic has developed a variety of additional methods (for instance, modal logic) to deal with more nuanced situations that are typical of real-world problems. Similarly, the so-called soft sciences are concerned largely with complex issues that require more sophisticated, but often less clear-cut, approaches; these approaches may be less satisfactory (but more realistic) than strong inference in that they yield probabilistic (as opposed to qualitative) answers. However, there is a second way of approaching the difference between hard and soft science that is more empirical in nature: instead of discussing and comparing the methods used by the two, let us look at their results and see whether the claim that, say, physics yields consistently more reliable results than, for instance, psychology is really true. This is what two social scientists, Larry Hedges and George Howard, set out to do in the late 1980s and early '90s. The results were rather surprising.

Soft vs. Hard Science: The Proof Is in the Data

One would think that scientists are by nature oriented more toward the facts than speculation, so it is rather surprising that there is much discussion, but little evidence about the difference between soft and hard sciences. One of the exceptions is a study published in 1987 by Larry Hedges, then in the Department of Education at the University of Chicago. Hedges began with a simple consideration: if it is true that physics (the queen of the hard sciences) "performs" much better than psychology (arguably the Cinderella of the soft sciences), one ought to be able to show, data in hands, that results from physics experiments are "better" than results from psychological experiments. But better in what sense?

Hedges thought that the difference between the two sciences should be evident in the "cumulativeness" of their results: physics should be making progress more steadily and at a faster pace than psychology. This is an important criterion, because lack of progress, i.e., lack of cumulative results over time, is one of the distinctive features of pseudoscience. For instance, the idea of intelligent design in biology—which we will examine in detail later in the book—has made no progress since its last serious articulation by natural theologian William Paley in 1802. Compare that to the stunning advances of evolutionary biology since Darwin's publication of *The Origin of Species* in 1859 and one has a good picture of the difference between science and pseudoscience.

As Hedges immediately recognized, however, cumulativeness in science can mean two very distinct, if related, things: on the one hand, one could examine theoretical cumulativeness, i.e., the advancement of a scientific field in terms of how well its theories account for how the world is. For instance, astronomy advanced by a giant leap when it abandoned the Ptolemaic, earth-centered, view of the universe in favor of the Copernican, sun-centered system. Then it made smaller but significant advances by realizing that the planets move in elliptical, not circular orbits (as Copernicus had incorrectly assumed); by discovering that the sun is only one star among billions in the Milky Way; and finally by placing our galaxy itself as only one of billions existing in the universe. Judging theoretical cumulativeness, however, is not simple, as it involves a degree of subjectivity, and—more crucially—it requires a long historical perspective. Psychology is a relatively novel science and it would therefore be rather unfair to compare its theoretical foundations with those of modern physics or astronomy, which have had a much longer history.

What then? Hedges opted for a more tractable measure of progress, focusing on empirical cumulativeness. The idea is that if physics, psychology, or any other science is really successful at describing the world as it is, then at a minimum its empirical results (from observation or experiment) ought to be consistent from one publication to another. To put it simply, if the earth is really round with a diameter of about 12,700 kilometers, the different methods to estimate its shape and size ought to yield approximately the same result. If it turns out that some measurement gives us 3,000 kilometers while others go up to as much as 100,000 then there is something seriously wrong with the way we do the measurements. Again, the comparison with pseudoscience is obvious: some creationists, for instance, believe that the earth is about 6,000 years old, while others accept the geological figure of 4 billion years, give or take. This is a staggering discrepancy of five orders of magnitude, which betrays the fact that creationists really have no idea of how old the earth is, or of how to measure it; in turn, this is yet another indication that creationism is no science.

So Hedges went about sifting the literature in particle physics (la creme de la creme in physics research) as well as in a variety of psychological fields, including studies of sex differences, of students' ratings of teaching, of the effect of racial desegregation programs, and others meant to provide a range from quasi-hard psychology (sex differences) to as-soft-as-it-gets research (desegregation studies). He used standard statistical tools in tabulating and comparing the results of a variety of studies published over a period of years in several specialized journals. The outcome was rather stunning: it turns out that the replicability of research findings in psychology (and therefore, presumably, the resulting empirical cumulativeness of that discipline) is no worse (or better) than the replicability of findings in particle physics. As Hedges put it: "What is surprising is that the research results in the physical sciences are not markedly more consistent than those in the social sciences. The notion that experiments in physics produce strikingly consistent . . . results is simply not supported by the data."[6]

Not only that. It turns out that some of the results in physics are much *less* reliable than one would think. For instance, Hedges compared the data obtained during two series of experiments aimed at estimating the mass of two fundamental particles: the electron and the proton. These are two of the best-known and best-studied particles, so one would expect a high degree of congruency among the outcomes of different experiments. Alas, this was not the case: experiments performed over a period of years (from the early 1960s through the mid-1970s) clearly show that the various estimates were not consistent with each other, and their so-called confidence intervals often did not overlap, meaning that the results were statistically significantly different from each other.

Data like this, of course, should not be interpreted to mean that physicists have no idea what the mass of the electron or the proton is. For one thing, we now have many more experiments, and their results are much more consistent. Moreover, it's not like the estimates reported by Hedges show the kind of huge variation that would make anyone seriously question fundamental aspects of nuclear physics. But the point remains that even the queen of science sometimes gets things wrong over a period of many years and that the quintessential example of soft science, psychology, actually displays a remarkable and surprising degree of consistency in its results. Should we conclude therefore that the difference between soft and hard science disappears when one looks at the data? Not so fast, says another social scientist, psychologist George Howard.

Howard published a paper in 1993 in *Theoretical and Philosophical Psychology* (a rather intriguing name for a scientific journal!) strangely entitled "When Psychology Looks like a 'Soft' Science, It's for Good Reason!"[7] The point of the article was to highlight the actual significance of Hedges's results: while the *reliability* of measurements may be roughly the same in physics as in psychology, the ability of the two disciplines to *explain and predict* is very different, and physics is unquestionably superior in that respect. Let us try to understand what this means, since it will lead us to some crucial insight into the distinction between soft and hard science, and later on between these and pseudoscience.

The important difference here is the one between estimating a quantity and accounting for the variation found around that quantity (in statistics these are called the mean and the variance, respectively). Let's say we are interested in finding out the mean height of the inhabitants of Brooklyn. It is not practical to go out and measure each and every one of them, so we take a sample: we randomly ask people to measure their height, and we record the results. If we do enough of these measurements, and we sample people truly at random (so that they are representative of the population at large), we should be able to come up with a reliable estimate of the mean height of the inhabitants of Brooklyn. By "reliable," I mean what Hedges meant when he was talking about empirical cumulativeness: if we repeat the "experiment," i.e., we go around Brooklyn a second and a third time, and so forth, and we repeat the procedure, we probably will obtain very similar estimates of the mean height of the people living there. In this, as Hedges showed, soft and hard sciences are equivalent: just like physicists can consistently get similar estimates for their quantities, so can psychologists, and presumably other scientists.

But now suppose we ask a physicist to explain the *variation* around his estimates of, say, the mass of the electron. After all, not every experiment will give the same results, as we have seen, and even within the same experiment he will not obtain exactly the same estimate every time he carries out his measurements. Here the answer of the physicist will likely be very different from that of the social scientist: the physicist will say that the variance around his mean value for the mass of the electron is due entirely to experimental *error*. In other words, the only possible cause for variation from one estimate to the other is the fact that human-made instruments aren't perfect, and that tiny variations in the experimental conditions are likely to alter the results. The social scientist, however, will say that experimental, or measurement, error is only one of the causes of variation in her sample: yes, it is possible that the experimenter has made mistakes in either carrying out the actual measurement or in recording it, but more important is the fact that there is natural variation among human beings for the trait of interest, height. People, unlike electrons, come in different sizes and shapes, and this variation in turn depends on a variety of factors, including their genetic makeup and the diet they have been exposed to while growing up. Moreover, there are differences between genders, and of course children are shorter than adults. All these differences *account for* (or, in statistical terms, *explain*) the observed variation, not just the experimental error.

According to Howard, the data clearly show that physicists (or, at least, particle physicists) can often account for 90 percent or more (in fact, close to 100 percent) of the variation in their samples, which is a very satisfactory result indeed. By contrast, social scientists (and organismal biologists) typically hit an explanatory "ceiling" around 25–35 percent of the variation, if they are lucky; i.e., the soft sciences usually cannot account for more than a quarter to a third of the factors that contribute to the variation in their experiments, on a good day. The rest of the variation is due to unknown factors that remain outside of the purview of the investigator. *That* certainly makes these results "softer" than those of the hard sciences.

The obvious question, then, is why is it that soft sciences typically have much less explanatory power (in the technical statistical sense of the term) than hard sciences? They are still doing better than pseudosciences, where the explanatory power is actually zero or close to it, but the difference is staggering and requires an explanation. Howard provides one, but I think it is only half correct: he claims that the lower performance of soft sciences is due to the higher number of causes involved in the sort of phenomena of interest to scientists in those disciplines, which we have seen was one of the explanations that Platt rejected a bit prematurely when addressing the same problem. But Howard also singles out one particular cause as different from anything encountered in other sciences, soft or not: human agency. He thinks that psychologists need to take into account that human beings have "free will," which they exercise when they make decisions, and which can throw a big wrench in standard scientific explanations of variation. A discussion of the very idea of free will would bring us far outside of the scope of this chapter (or, indeed, this book).[8] Suffice to say that free will is a way to label the complex decision-making processes in which the human brain engages, consciously as well as probably subconsciously. There is no reason to consider this as a unique kind of causal explanation, and it makes more sense to simply treat it as one of the many interacting causes that underlie any complex natural process, human or not. Seen this way, the reason there is not much difference in explanatory power between, say, psychology and evolutionary biology has nothing to do with the human element per se, but is due to the fact that both disciplines—unlike particle physics—deal with complex layers of causality.

Let us take stock of what we've got so far. While Platt's argument that strong inference is very effective (when it can be applied) is right, he was wrong in the conclusion that what makes a soft science soft is the inability or unwillingness of its practitioners to employ that method. Hedges's work showed that the perception of a much higher level of reliability of results in physics compared to psychology is not, in fact, borne out by the empirical evidence. But Howard reminded us that the ability of physics to explain its results is much higher than that of psychology, even though I think he unnecessarily singled out human agency as the explanation for the difference. Next we turn to a more recent and sophisticated view of the same problem. We are about to learn that not only are there good reasons (other than some scientists' laziness) why the results of the soft vs. the hard sciences differ in quality, but that these reasons have to do with the very nature of one of the most mysterious concepts in all of science: causality.

Causal Asymmetry and the Difference between Predicting the Future and Explaining the Past

"[Historical hypotheses] can never be tested by experiment, and so they are unscientific . . . No science can ever be historical." This absurd opinion was expressed by none other than Henry Gee, an editor at one of the most prestigious scientific journals in the world, *Nature*.[9] If Gee were right, large parts of evolutionary biology and geology, as well as the whole of paleontology, including human paleontology, and astronomy, to mention but a few, would be excluded from science curricula; whole exhibits at the American Museum of Natural History in New York (such as the Hall of Human Evolution) would have to be closed; and many of *Nature*'s own papers would have to be retracted as nonscientific. Fortunately, Gee was spectacularly wrong, so there is no need for an overhaul of how we carry out scientific research or how we present it to the public.

Nonetheless, Gee's remarks offered the opportunity to philosopher Carol Cleland to write one of the most lucid commentaries on the scientific method that I have read,[10] and one on which I will draw in this section to further explore the idea that there is more than one kind of science and more than one kind of scientific method.

To begin with, let us clear the field from a common misunderstanding about what it means to do science, a misconception that is unfortunately widespread even among scientists: one does not need experiments to do science. While this claim may sound strange and counterintuitive at first, a moment's reflection will show that it is obviously true: astronomers do not conduct experiments, and yet we think of astronomy as solidly situated within the sciences, not the humanities or the pseudosciences. Why? Because astronomers can carry out the two fundamental activities that, jointly considered, truly characterize a science: systematic observations and the construction and testing of hypotheses. For instance, one of the stunning achievements of astronomy (or, more properly, of one of its branches, cosmology) during the twentieth century was the confirmation that the universe really did have a beginning, a "Big Bang" that started a series of processes that led to the world as we know it. But how can one test this idea? How do we know that the universe hasn't always existed, an alternative theory actually proposed by cosmologists Hermann Bondi, Thomas Gold, and Fred Hoyle in 1948 and known as the "steady state" theory of the universe? One of the crucial pieces of evidence in favor of the Big Bang over the steady state theory was discovered in 1964 by Arno Penzias and Robert Wilson, two radioastronomers working at Bell Laboratories. Penzias and Wilson found that the universe is permeated by a low-level background radiation, characterized by a temperature of about three degrees Kelvin (i.e., three above absolute zero, or –454°F). Such background radiation is precisely what the Big Bang theory predicted we should observe as a result of the initial explosion, and is of exactly the magnitude predicted by the theory: Penzias and Wilson (who later won the Nobel for their discovery) had found the mother of all smoking guns! There are several other pieces of circumstantial evidence supporting the Big Bang theory, all pointing to the same conclusion: the party started with a blast a bit less than 14 billion years ago. Still, the finding of a clear smoking gun, according to Cleland, is one of the reasons that make historical sciences like astronomy perfectly good members of the scientific enterprise, no apologies needed just because its practitioners cannot do experiments.

Take another famous example: the theory that the dinosaurs got wiped out (or, at least, finished off, since they were already in decline at the time) by a large meteor that hit the earth 65 million years ago, marking the line that divides the Cretaceous and Tertiary geological periods (the so-called K-T boundary). The impact theory had been around for some time in a variety of forms and had coexisted with alternative explanations until 1980. That year, a team of researchers led by the father-son pair Luis and Walter Alvarez (the first a physicist, the second a geologist) discovered an unusual layer of iridium in rocks dated to the K-T boundary. The discovery was important for two reasons: first, the iridium was present worldwide, so it was not the result of a local geological event; second, and most tellingly, iridium is a very rare material on earth, but it is abundant in a particular kind of meteorite. This discovery spectacularly confirmed the idea that a large body had indeed struck the earth 65 million years ago, lending strong support to the theory that a cosmic catastrophe had been at least partially responsible for the demise not just of the dinosaurs, but of a great percentage of life forms existing on the planet at the time. Still, the problem was that by the calculations of Luis and Walter Alvarez, the impact should have formed a crater of about 250 kilometers in diameter: where was it? The crater—another smoking gun—was discovered in 1990 near Chicxulub, off the coast of the Yucatan peninsula in Mexico. Not only was the newly discovered crater of the right size, but geologists were able to determine from its shape the angle of impact of the meteor; this was such that the blow would have created giant tsunamis that would have hit the coast with devastating impact. Sure enough, once they knew what to look for and where, scientists found ample evidence of tsunami activity in the area, going back to just around the time of the K-T boundary. If this is not good science, pace Gee, I don't know what is!

These and similar examples are easy enough to uncover, and they make two crucial points: first, good science does not require experiments, it can be done with an intelligent use of observational evidence; second, there is more than one way to do science, depending on the nature of the questions and the methods typical of the field.

Cleland made one more salient contribution to the debate on the nature of science and the difference between historical and nonhistorical sciences (roughly, though not exactly, what we have been calling "soft" and "hard" sciences, respectively). She focused her attention on what is known in philosophy of science as the "asymmetry of overdetermination," an esoteric name for a very important concept that we need to explore and make sense of. The basic idea is simple enough: the relationship between the present and the past, when it comes to scientific investigation, is dramatically different from the relationship between the present and the future. What this means is that scientists who try to understand the past (astronomers, paleontologists, evolutionary biologists) find themselves dealing with a very different kind of problem from those facing scientists whose main business is to predict future events, including the outcomes of experiments (physicists, chemists, not to mention atmospheric scientists).

Let's use a couple of Cleland's everyday examples to understand what the asymmetry of overdetermination means, and the relevance to actual scientific practice will become obvious. We often hear in movies, TV shows, and crime detective novels that "there is no such thing as a perfect crime." It turns out that there are some solid reasons behind this intuition: when someone commits a crime, no matter how careful the perpetrator is, there will be a number of "traces" that might help an acute investigator to reconstruct enough of what happened to identify the culprit. There will be fingerprints or hair or skin fragments or blood; there will be credit card transactions, perhaps, or phone bills related to something the perpetrator did in connection with the crime; there will be witnesses, if not of the crime itself, then of some of the preparatory steps that led to it; and so on. It is exceedingly difficult to go back and erase all of the historical traces of what happened, and it will take only a few of these for a detective to solve the crime. This, of course, does not mean that every crime goes solved, because human investigators are fallible, and because historical traces decay with time: the more time passes, the more difficult it is to put together what happened (that is why so called cold cases are much more difficult to solve). The point is that it is feasible, and in some cases indeed downright easy, on the basis of a few clues, to reconstruct a crime. On the other hand, it is next to impossible to *predict* when and how a crime will be committed. This is because the same high number of variables that allow the solution of the crime after it has happened must come into play in a specific order for the crime to be actually committed: someone has to have a motive, plan the details of the deed, act to set the conditions right, and then actually do it. If any one of these steps is halted or goes wrong, chances are that no crime will occur.

Another example provided by Cleland makes the same point perhaps even more clearly: consider a house fire. Once it has happened, it is pretty easy to see that it did happen and probably even to infer what caused it. This is because, again, the number of clues, or traces, will usually be high. Now suppose the fire department determines that the fire was caused by a short in the electrical circuit. This may be technically true in that specific instance, but in effect the number of causes is very high: they include the very existence of an electrical circuit; the fact that the house is made of wood, say, rather than bricks; the failure of the sprinkle system to work and extinguish the fire; the absence of people inside the house at that moment, who could otherwise have extinguished the fire manually or asked for early help; and so on. Any one of these factors might have resulted in the fire never happening, or at least in the house not being destroyed. So, again, while it is relatively straightforward to figure out what happened after the fact, it is mightily difficult to predict what will happen and when.

Now that we know about the asymmetry of causal determination (past < present vs. present > future) we are in a better position to appreciate why, for instance, paleontologists can find—given the right circumstances—the smoking gun for events like the meteorite that created the K-T boundary, but any scientist, no matter what the discipline, would have a very difficult task predicting whether, when, and where such an impact might occur again. This being the case, how is it possible that some sciences, such as particle physics, seem to be able to make astoundingly precise predictions about future events, beginning with the outcomes of their own experiments? The answer is that in most cases the type of experiment that makes such predictions possible, say in quantum mechanics, is carried out under highly controlled and very special conditions in the laboratory. In those circumstances, the scientist has almost complete control of the already very few causal factors on which the result of the experiment depends. Indeed, even under these ideal conditions physicists and chemists often have to repeat their experiments many times, taking care to control for variables such as the exact specs of the experimental apparatus or even the physical layout of the lab in which the measurements are being undertaken. Despite all this, we have seen earlier that different labs can still come out with wildly diverging estimates for basic physical quantities, such as the mass of the proton.

What Cleland's analysis makes clear is that hard sciences are not "better" than soft sciences, but rather that they are characterized by different kinds of problems, and that each science has its own methods and accompanying limitations. Historical sciences have the advantage that even minute traces of past events may be sufficient to test a hypothesis, especially if a smoking gun is found. On the other hand, they perform abysmally when it comes to predicting future events because of the rapidly diverging number of possible outcomes that arises from even a small number of causal factors at play. The experimental sciences do perform incredibly well when it comes to measuring quantities with very high precision, or when they make predictions about the behavior of systems during a controlled experiment. But the level of accuracy drops dramatically once the number of variables increases, especially if one moves outside of the lab environment and into the real world, where the situation easily becomes messy and conditions are difficult or impossible to standardize.

I think it is fair to conclude from what we have seen so far that science is a heterogenous group of disciplines that carry out research in the natural world in distinct ways. The common thread in all science is the ability to produce and test hypotheses based on systematically collected empirical data (via experiments *or* observations). How these hypotheses are generated, how exactly scientists go about testing them, and the degree of success we can expect from different sciences varies from science to science and from problem to problem. This variation is not random, however, and it depends largely on two factors that we have discussed at some length in this chapter: on the one hand, we have a continuum from completely historical (paleontology, astronomy) to partially historical (evolutionary biology, geology) to essentially ahistorical sciences (physics, chemistry). The more historical a discipline, the more its methods take advantage of the "smoking gun" approach that we have seen working so well with the extinction of the dinosaurs and the beginning of the universe. The more ahistorical a science, the more it can produce highly reliable predictions about the behavior of its objects of study, though these predictions are largely confined to highly controlled conditions in the laboratory.

On the other hand, we have a second continuum, from sciences that deal with simple, highly tractable systems where one can apply strong inferential methods (physics, chemistry) to sciences dealing with extremely complex objects, where statistical treatment is necessary and where the ability to explain and predict phenomena is much reduced (evolutionary biology, psychology). All these differences have nothing to do with the quality of the intellectual endeavors themselves and even less to do with the intelligence or brilliance of the participating scientists. We will now move to the even trickier consideration of a number of disciplines that definitely bring us out of mainstream science, some of them looking a lot like sciences and yet raising serious questions about the empirical testability of their theories. Beyond that, we will shift even further away from science proper and descend to explore some of the depths of actual pseudoscience.

CHAPTER TWO

Almost Science

Any fool can make history, but it takes a genius to write it.— Oscar Wilde

Sometimes I think the surest sign that intelligent life exists elsewhere in the universe is that none of it has tried to contact us.—Bill Watterson, cartoonist, *Calvin and Hobbes*

We began our exploration of the nature of science by examining the idea that there is a simple distinction between "hard" and "soft" science, and found the idea wanting. We learned that there is indeed a difference there, but that it is much more nuanced than often portrayed, depending on a variety of factors that include the complexity of the subjects of study, the social availability of funding for those studies, and even how cool (or not) college students may perceive a certain academic career to be. Here we will begin to move away from those areas of research that few would question *are* in fact science, soft as they may be, and toward a middle land of quasi-science populated by strange animals like evolutionary psychology, history, and even the search for extraterrestrial intelligence (SETI).

There are several reasons why this middle territory between science and nonscience is interesting. First and most obviously, the disciplines that occupy it address questions about which we care either because they affect our lives (the evolutionary roots of psychology) or because they strike at some of our fundamental curiosities about the universe (SETI) or about ourselves (history). For the purposes of this book, however, the middle land that I call "almost science" is also bound to tell us something about the nature of science itself and how it may differ from pseudoscience. The disciplines in middle land may one day be recognized as full members of the scientific enterprise, as is happening to areas of psychology that are turning into cognitive science; or they may slide into pseudoscience, as happened in the past to astrology and parapsychology; or they may remain poised at the borderlands, perhaps to be joined by altogether new fields of speculation and inquiry. We will start with a field that one would hardly suspect of harboring a possibly nonscientific quest: physics itself, the queen of the hard sciences.

Is Physics Turning into Philosophy? On String Theory and the Multiple Interpretations of Quantum Mechanics

By most people's account, physics is the queen of the sciences. Philosophy, by converse, is, according to some physicists (for example Steven Weinberg, in his *Dreams of a Final Theory*[1]), a useless, perhaps even dangerous enterprise because it can slow scientific progress. Yet some authors have argued that physics is in danger of becoming "just" philosophy, or worse, indistinguishable from nonscientific notions such as the anthropic principle, the idea that the universe is fine tuned to allow for the appearance of life (a close cousin of so-called intelligent design theory, which we will discuss later on).

George Ellis, for instance, reviewing Leonard Susskind's book *The Cosmic Landscape*,[2] concluded that "heavyweight physicists" are claiming to prove the existence of parallel universes "even though there is no chance of observing them." Along similar lines, Michael Atiyah, commenting on another book about theoretical physics, Lawrence Krauss's *Hiding in the Mirror*,[3] observed that there is a danger of a "mathematical take-over" of physics, leading to speculations that, while mathematically elegant, are "far removed, or even alien to, physical reality."

These are harsh words, no less damning of those that Platt used in 1964 to extol the virtues of physics over "softer" sciences, as we have seen in the last chapter. What is going on here? According to Ellis, currently fashionable ideas in physics and cosmology, such as the multiverse or string theories, are characteristic of a discipline where "faith in theory tends to trump evidence." Take string theory, for example: it is a mathematically beautiful construct that has the potential of unifying general relativity and quantum mechanics. The problem is that—at the moment—it is also entirely untestable empirically, which makes it an interesting philosophical position, but hardly science. More on this in a moment.

Let's start by considering the multiverse theory, an interpretation of quantum mechanics, itself arguably the most successful of all scientific theories ever proposed (in the sense that its experimental predictions are regularly verified to a high degree of accuracy, though remember our discussion of "accuracy" in the last chapter). Now, why would a scientific theory need an "interpretation," a word that is usually reserved for sacred texts such as the Old Testament? In the summer of 2008 I witnessed a fascinating panel discussion at New York University in which two theoretical physicists, one experimentalist (i.e., one of those people who actually do what it takes to test theories), and even one philosopher were excitedly debating which interpretation of quantum mechanics makes more sense (the fact that the event was being held in front of a full house of several hundred paying people simply gives you a measure of what sort of bizarre place Greenwich Village in New York City really is). The two theoretical physicists participating in the panel were Max Tegmark, a cosmologist at MIT, and Brian Greene, a professor of physics at Columbia. The experimentalist was William Phillips from the University of Maryland, who won the Nobel for his work on manipulating atoms using lasers. The philosopher was former physicist David Albert, also at Columbia. What was the discussion all about?

The four participants easily agreed that the mathematics that is at the foundation of quantum mechanics is sound, and moreover that it "delivers," meaning that it makes experimentally verifiable predictions. This sounds like solid hard science, so one wonders why the fuss. Indeed, Phillips claimed that there shouldn't be any fuss: he subscribes to a school of thought that could be summarized as "shut up and calculate" (his words). The idea is that if the mathematics is correct, and if it is helpful in doing experiments, that is all one ought to care about, forget trying to "interpret" the meaning of the theory itself. Not so fast, responded, perhaps predictably, the philosopher. Albert argued that science—unlike, say, engineering or medicine—isn't just about what works and what doesn't. Scientists want to make sense of the world as it is, not just invent mathematical formalisms that account for its behavior. It's a good point, and a simple example will drive it home: consider exactly why Copernicus and Galileo got into trouble with the Catholic Church for their insistence on a heliocentric model of the solar system. If they had limited themselves to say that the new theory was just a mathematical device to better calculate the positions of the planets (as stated in the preface of Copernicus's book), the Inquisition would hardly have been so keen to intervene. But the dangerous (for the religious authorities), revolutionary aspect of the theory was that it claimed that the sun *really* is at the center of the solar system, and that the earth and the other planets *really* rotate around it. The relevant question therefore, said Albert during the discussion at NYU that night, is what do we think quantum mechanics tells us about what the microscopic world is *really* made of?

So what did the two theoretical physicists think of that deceptively simple question? Tegmark goes for a quasi-mainstream view of quantum mechanics known as the many-worlds interpretation; Greene prefers a more recent answer that purports to present a whole new view of the universe, the above mentioned string theory; and both disagree with one of the classic renditions of quantum mechanics known as the Copenhagen interpretation. These are by far not the only competitors currently jousting for dominance in theoretical physics,[4] but a brief look at each will give us some flavor of why the hardest of the hard sciences is at the moment poised on the brink between pure speculation and abstract mathematical or philosophical inquiry.

Let us start with some of the facts, on which all quantum physicists agree, regardless of their school of interpretation. Before quantum mechanics, physicists thought that fundamental components of reality, such as electrons (which, with their movement, make possible the very useful phenomenon of electricity) or photons (the constituents of light), are either particles or waves. If they are particles, they behave in a certain way, for instance they hit targets in a specific spot if fired by an experimental apparatus. If they are waves, throwing them against a target (a screen, say) will generate a diffuse pattern, depending on the characteristics of the wave (i.e., its frequency and amplitude). To see the difference, just think of equivalent situations in the macroscopic world: firing a bullet against a wall will result in a very different outcome from generating a wave of water against the same wall.

The problem is that a famous experiment known as the double-slit experiment painfully showed that the quantum world doesn't behave at all like the familiar macroscopic one. The experiment was originally performed back in 1801 by Thomas Young, with the aim of settling the dispute of whether light is made of particles, as argued by Newton, or is a wave similar to sound, as suggested by Christiaan Huygens and Augustin-Jean Fresnel. The full import of the experiment was not appreciated until the advent of quantum mechanics at the beginning of the twentieth century, and physicist Richard Feynman used to say that all of quantum mechanics can be understood from a careful consideration of what the double-slit experiment means (this, I believe, is a slight exaggeration, to say the least). Here is how it works. Suppose you direct a source of photons (though electrons will do too) toward a barrier with a single slit, behind which there is a screen. What happens is that a number of photons will pass through the slit and hit the target, creating a simple pattern consistent with the idea that photons are particles. Now change the setup slightly and use a barrier with two slits: common sense predicts that you should see photons passing through both slits and hitting two clearly separate points on the screen, but you don't! What happens instead is that the screen shows the complex pattern characteristic of waves interfering with each other. How is this possible? The straightforward conclusion is that photons (or electrons, or a number of other particles) behave as particles if you use one slit, but *switch* to a wavelike behavior if you use two slits! This seems to fly in the face of common sense, though the problem, of course, may be that common sense is a really bad guide to interpret results from subnuclear physics, in which case so much the worse for common sense.

Physicists typically say that the double-slit experiment is a demonstration of the *dual* nature of light: wave *and* particle, depending on the circumstances. But this doesn't seem to help our intuition of what is going on at all: how can the same thing be a particle or a wave depending on how it is observed by humans? Mars is a planet, not a star, regardless of which telescope one uses to observe it. Why is this nice and comfortable consistency of macroscopic objects lost in the quantum world? Physicists make sense of results like those from the double-slit experiment in this way: in reality—more on this concept in a moment—photons are characterized by a "wave function," a mathematical construct that tells you what the probability of finding the photon in a particular location is. When you actually make the measurement (i.e., when you stop the movement of the photon by interposing the screen), this very action "collapses" the wave function to a single value; that value, no longer a probability, tells you the position of the photon at the moment of the measurement. In a very deep sense, it seems, it really is the interference of the human observer that "creates" the reality of quantum experiments, because the wave function does not collapse (and hence the photon does not have a definite position) *until the measurement is carried out.*

We can now begin to appreciate why the mathematics (indubitably correct) and experimental results (unquestionably accurate) of quantum mechanics nonetheless seem to beg some sort of "interpretation" of what is *really* going on. It is therefore time to consider the three main ideas debated at the NYU panel, to see how physicists make sense of apparently bizarre constructs such as the wave function. Let's begin with the Copenhagen interpretation, which goes back to the early years of quantum mechanics and specifically the work of Niels Bohr and Werner Heisenberg in Copenhagen (hence the name of the school) during the 1920s. Bohr in particular was influenced by a then prevalent school of philosophy known as positivism that asserted that science has to concern itself only with what can be observed or measured. This led Bohr to the rather counterintuitive notion that one should accept the results of the experiments and of the mathematical calculations, but that not much more could be said about the underlying reality of things, a position not too far removed from the above mentioned "shut up and calculate" stand taken by Phillips. Some adherents to the Copenhagen school actually maintained that wave functions are not just mathematical abstractions, but are in some deep sense "real" properties of physical systems, although it is anybody's guess in what sense a probability distribution can be physically real. Others, when pressed, admitted that perhaps the very concept of reality is overstated (apparently Bohr himself was sympathetic to this view).

If you find all of this a bit fuzzy and unsatisfactory, you are in good company: Albert Einstein positively disliked some of the perceived implications of quantum mechanics and the Copenhagen school in particular, especially the idea that the universe is fundamentally governed by random events (he famously quipped that "God doesn't play dice," to which Bohr allegedly replied that he wished Einstein would stop telling God what to do). Indeed, Einstein spent much of the time during the last part of his career trying to come up with a more solid interpretation of quantum mechanics, one that would recover some sense of physical reality beyond randomness and probabilities (he failed). Other physicists eventually produced one such attempt, the so-called many-worlds interpretation endorsed by Tegmark at the discussion I enjoyed at NYU. The many-worlds theory basically maintains that the reason quantum mechanics seems so strange is because we have access to only one of an infinite number of worlds. From our narrow perspective, the output of certain measurements (like that of the double-slit experiment) seems random and probabilistic, but that is an artifact of the fact that, literally, we don't have the full picture. In fact, proponents of the many-worlds hypothesis say, reality is such that there are infinite universes, all differing to a greater or lesser extent from each other. Because of this infinity of possibilities, everything that can happen does in fact happen, just not in our little corner of the "multiverse," the only one to which we have access. This may sound preposterously *Star Trek*–like, but bear in mind that it is a mainstream view among fundamental physicists.

Not so mainstream, and yet increasingly popular, is the third idea we will consider briefly: string theory, supported by Brian Greene at the NYU event and beautifully explained in his book *The Elegant Universe*.[5] This is where things have the potential to get even further remote from science proper, despite the fact that people working on string theory are unquestionably legitimate scientists doing some of the most cutting-edge work in their discipline. In fact, it is *so* cutting edge that some consider it to be beyond the line that separates science from nonscience. First we need to understand why people are attempting to construct a new theory of the physical sciences at the same time that everyone acknowledges the spectacular successes of the two currently dominant theories in physics, quantum mechanics and general relativity. The problem is that these two theories work very well when applied to their proper domains, respectively, the very microscopic and the very macroscopic worlds. That is, quantum mechanics does a beautiful job at predicting how matter behaves at the scale of quarks, electrons, photons, and the like. Relativity, on the other hand, works very nicely when it is a question of describing the behavior of systems like planets, galaxies, and so forth. However, when one tries to apply quantum mechanics to typically relativistic problems—such as developing a quantum theory of gravity—the two theories behave very differently and their predictions diverge. The same happens if one attempts to apply relativity to the subnuclear world. The logical conclusion is that either one of the two theories is seriously flawed, or that they both are, or at the very least that they are incomplete. In any case, it seems clear to physicists today that there is something amiss, hence the search for a so-called final theory of everything, a rather boastful name for a framework that would unify our understanding of the basic forces of the universe and describe all the particles of which matter is made.

Enter string theory. It is based on the idea that the known subatomic particles are not the most fundamental constituents of matter. Rather, the stuff of which everything is made are even stranger objects termed "strings," which take the properties of various particles/waves depending on the frequency at which the strings vibrate. This notion has a variety of strange consequences, including that the universe actually possesses eleven, not just four dimensions (the three familiar spatial dimensions plus time). Another consequence is that if string theory (which is actually a family of related theories differing in some of their details) is correct, then physicists would find themselves with a mathematically internally coherent quantum theory of gravity, something they have been after now for several decades.

String theory, if you think about it, is a beautiful idea: it means that everything in the universe is made of just one kind of thing and that this fundamental aspect of matter takes a variety (not an infinite variety, mind you) of forms, depending on a simple property such as its vibration frequency. It is so elegant an idea that it deserves to be true. But is it? Therein lies the problem: at the moment, at least, string theory does not seem to make any empirically testable predictions that both differ from those of other competing theories and that can conceivably be evaluated in actual experiments. If one defines science as a type of inquiry into nature characterized by the availability of empirically testable hypotheses, then string theory is, strictly speaking, not science.

Perhaps, then, we should think of string theory, the multiverse interpretation of quantum mechanics, and similar intellectual pursuits as more akin to philosophical investigations, albeit mathematically rigorous and informed by the empirical science. Of course, if one begins with the common assumption that physics is not only the queen of the sciences, but that science is in turn the sovereign of all intellectual disciplines, then it is an insufferable insult for string theory to be "downgraded" to mere philosophy (or even to purely mathematical work). Then again, arguably this peculiar relationship between science and philosophy is nothing new. Philosophy has often been the placeholder for areas of intellectual inquiry that have subsequently moved to the domain of science. Physics itself, up until René Descartes, was part of natural philosophy. Descartes considered himself a scientist, not a philosopher (which is the way he is remembered today), and it is obvious only with hindsight that he lived at the time of separation between the two modes of inquiry. Similarly, psychology originally was a branch of philosophy, and philosophy of mind is now increasingly turning its attention to neurobiology and cognitive science (with the result that the latter two may eventually replace the former). The role of philosophy in each of these cases was to explore the intellectual terrain and clear up the conceptual mud until ideas and techniques were mature enough to permit a shift to empirically based inquiry. Perhaps multiverses and strings are playing the same mud-clearing role, in preparation for whatever further advancement about the fundamental nature of reality science will bring next.

Obviously, this is not to argue that all areas of logical and philosophical analysis will eventually transform themselves into science, regardless of how much time or technological advance one allows. For example, despite the fact that biology provides some insights into human concepts of morality and aesthetics,[6] it is hard to imagine simply turning the latter two disciplines over to science departments, because they involve human values, not just matters of fact. Moreover, it may simply well be the case that we will never find a way to test the idea that there are multiple universes out there. Just because we are curious animals, there is no assurance that nature behaves in a way that allows us to get answers to every mystery that happens to intrigue us.

Another way to look at the conundrum presented by some theories currently being entertained by physicists is that there is a distinction between logical and empirical possibility, and—correspondingly—between logical and empirical truth. The realm of logical/mathematical possibility (mathematics may be considered a branch of logic, though this is by far not a universal position) is much vaster than that of empirical possibility. For example, it is logically possible for a substance made of two atoms of hydrogen and one of oxygen to behave very differently from water, if the laws of physics were just slightly different from what they are. Yet, once the laws of our universe are fixed, certain facts become empirically impossible. There are other things that are both logically and empirically possible, but have not been realized—for example the existence of unicorns. At the opposite extreme, if something is logically impossible (for example, for me to be and not to be myself), then it is a fortiori empirically impossible.

Strings and multiple universes are certainly logical possibilities, since they are features of mathematical theories (assuming that the math doesn't contain mistakes). What we don't know is whether they are also empirically possible and, if so, whether they have in fact been realized. It is the search for an answer to the latter two questions that distinguishes science from logic-mathematics (and philosophy). Neither string theory nor the possibility of multiverses have yet crossed the conceptual line separating logical from empirical possibility, but of course that does not turn them into pseudosciences, it just tells us that science is characterized by a fuzzy borderline with other types of inquiry that may or may not one day become sciences. The next example of this type strikes even some physicists as more closely related to science fiction than to good science, and yet it is a source of endless fascination for most of us: the search for extraterrestrial intelligence.

SETI: How Do We Find Out If Anybody Is Out There?

Arguably, people have been wondering about life in the cosmos outside Earth ever since the human species has been able to wonder about anything at all. Even worshiping stars and planets as gods can be seen as a way to project agency, and therefore intelligence, outside of the narrow confines of our own home among the stars. Until the twentieth century, however, the question of extraterrestrial intelligence was a matter of philosophical debate and speculation. SETI, as the quest is now so often referred to, entered the scientific discourse in 1959, when Giuseppe Cocconi and Philip Morrison, two physicists at Cornell University, published a paper in *Nature* arguing that human technology was now advanced enough to actively search for radio signals emitted by other civilizations in the galaxy. The following year, radio astronomer Frank Drake started the short-lived but historically important project OzMa, during which he used the radiotelescope at Green Bank in West Virginia to actually listen and ascertain whether anyone was trying to communicate.

Project OzMa did not discover any sign of extraterrestrial intelligence, and neither have any of its successors so far, including extensive searches conducted in the Soviet Union during the cold war, NASA's Project Cyclops, NASA's Project Phoenix, and University of California, Berkeley's Project SERENDIP, among others. There are currently several efforts along similar lines, the most ambitious of which is the search undertaken using the newly constructed (in fact, still under construction at the time of this writing) Allan Telescope Array, the first group of radiotelescopes expressly dedicated to a SETI project. Moreover, millions of people are now contributing to SETI by virtue of the ingenious "SETI at Home" project, which distributes software that anyone can acquire for free[7] and run on any personal computer. The software periodically downloads data from the SETI project on your machine, works in the background (or as a screensaver) to analyze it, and then sends it back to the team at Berkeley, where processed data from all over the world are compared, searching for a radio signal that cannot be attributed to natural phenomena. Hundreds of millions of dollars, decades of active research by several teams of scientists, and millions of hours of computer time have been spent. So far, however, nothing has turned up. Is this science?

SETI certainly does meet the minimum criteria for science that we have identified so far: it is based on the ability to conduct systematic observations, often of particularly promising star systems and at radio frequencies deemed more likely to be used for interstellar communication; and it allows for the possibility of empirical confirmation of its central hypothesis (namely, that there is intelligent life out there). The fact that such a hypothesis has so far failed to be supported, however, puts SETI in a different category from ordinary science, because there is no reasonable way to *disprove* the hypothesis either, given that the search is more akin to looking for a needle in a proverbial (and huge!) haystack. We have seen at the beginning of this book that falsification, the ability of a hypothesis to be rejected, at least potentially, is not necessarily the all-encompassing solution to the demarcation problem that Popper thought it was, because confirmation also plays an important role in science. As a result, hypotheses are usually given a fair chance to succeed despite initially negative results. But in the case of SETI, negative results are what are expected most of the time, perhaps even forever, regardless of the truth of the central hypothesis. This raises the question: when will SETI researchers think that enough negatives have been accumulated to reject the hypothesis of existence of other technological civilizations? If the answer is that such hypothesis can never be rejected, regardless of the empirical results, that pushes SETI uncomfortably close to the status of pseudoscience.

There is another way to look at the problem, based on an additional element (besides empirical evidence and testability) that needs to be present for a research activity to qualify as scientific: we also need a theory. This means that we need to ask what is the conceptual underpinning of the SETI program, something that makes it more than the simple hope that we will find someone out there, eventually. This is a rather complex question, which we are going to tackle only briefly, but it turns out that the SETI community does recognize a piece of theoretical work as having been pivotal for discussions within the field. However, there is also a crucial, some would say fatal, theoretical objection to the whole enterprise, possibly leading to the suspicion that SETI, if not really a pseudoscience, is going to be one big waste of time and resources. It is then time to consider both the famous Drake equation and the infamous Fermi Paradox.

The Drake equation gets its name from the same Frank Drake who pioneered the empirical SETI work with his Project OzMa in 1960 (and who still, incidentally, is actively involved with the main SETI project). Drake wrote out the equation originally as a way to organize thoughts for an upcoming conference on SETI, but the equation soon came to be considered a substantive theoretical principle behind the search for extraterrestrial intelligence. The equation is this:

$$N = R^* \times f_p \times n_e \times f_l \times f_i \times f_c \times L$$

where the left-hand term, N, is the number of civilizations in our galaxy that are at a technological stage that allows for interstellar communication. The right-hand terms are: R^*, the rate of formations of stars in the galaxy, on average; f_p, the fraction of stars that have planets; n_e, the number of planets capable of supporting life orbiting around stars that do have planetary systems; f_l, the fraction of planets capable of supporting life where life in fact does evolve; f_i, the fraction of the latter that develops intelligent life; f_c, the fraction of civilizations capable of producing interstellar signals; and L, the length of time during which a technological civilization exists and is capable of sending signals through space.

The problem with the Drake equation becomes clear after a careful look at the variables that define it: most (some would say all) are very difficult or downright impossible to actually estimate, which in turns means that N, the output of the equation, can take pretty much any value, depending on the personal preferences of the investigator who uses it. For instance, Drake himself plugged in the following numbers, back in 1961:

$$R^* = 10/\text{year};\ f_p = 0.5;\ n_e = 2;\ f_l = 1;\ f_i = 0.01;\ f_c = 0.01;\ L = 10{,}000 \text{ years}$$

Based on these guesses, Drake's estimate for the number of civilizations capable of communicating with us in the galaxy was N = 10. This would make for a surprisingly thin galactic yellow pages, if one considers that the Milky Way numbers around 100 billion stars, but it might still be reason enough to mount a SETI program.

Since 1961 astronomy has made some progress in areas that are related to the Drake equation, especially in estimates of the rate of star formation, and most spectacularly in the actual discovery of a number of planetary systems outside our own. Here is what the Drake variables look like at the moment, based on NASA figures and a variety of papers from the primary literature in physics and astronomy:

$$R^* = 7/\text{year};\ f_p = 0.2\text{–}0.6;\ n_e > 0.005;\ f_l > 0.13;\ f_i = 1.0 \times 10^{-7};\ f_c = 0.01;\ L = 10{,}000 \text{ years}$$

With these numbers, or rather range of numbers, plugged in, we get N = 2.3, which actually makes the already lowball estimate by Drake look quite optimistic. You can play the game yourself on the SETI Web site and plug in whatever numbers you like in the Drake equation to see how different choices affect the estimate of N.[8]

Let us step back for a moment and reexamine each of the terms in the Drake equation, not from the point of view of what might be the best current estimate, but of whether they are estimable at all; if they are not, then the whole idea of using the Drake equation as a theoretical underpinning for SETI runs into serious trouble. R^*, the rate of star formation in the galaxy, is something astronomers have methods to estimate, at least approximately. Although one has to make assumptions about how constant (or not) the rate is over the lifespan of the galaxy, this is no different from estimating any uncertain parameter in other branches of science (say, for instance, the rate of production of new species in biology). The fraction of stars with planets, f_p, was out of reach of any empirical investigation when Drake first proposed his equation, but now that astronomers have found reliable ways to discover new planets (more than three hundred so far, with a new NASA mission, *Kepler*, which beginning in 2009 has been dedicated expressly to the search of Earthlike planets) we begin to have reliable data here too.

The trouble starts when we move on to consider the third variable in the equation, n_e, the number of planets capable of supporting life. To begin with, of course, it depends on what one means by "life." The unspoken assumption is that we are talking about carbon-based life forms similar to those that evolved on Earth. But one can reasonably object that there may be other types of life out there, some based on different sorts of chemistry (often silicon-based life forms are brought up as possibilities). Of course, this should boost the morale of SETI researchers, because it would mean a higher likelihood of life forms evolving on other planets. But it also undermines the usefulness of the Drake equation, because we have no idea even *if* noncarbon life is possible, let alone how to go about estimating its probability. The second major problem with guessing n_e is that we really only have one known example of life-supporting planets: our own. With a sample size of one it is extremely bold of a SETI researcher to extrapolate to the entire galaxy. It is true that one can resort to what statisticians refer to as the principle of mediocrity: if you only have one example of a given phenomenon, the most sensible thing to do is treat that example as representative for that class of phenomena. Again, this is good news for SETI, because it implies that life-supporting planets are relatively frequent, but this support comes at the cost of an almost ludicrously low level of credibility for the estimate itself.

The same sort of problematic considerations apply to the remainder of the terms in the Drake equation: we have no idea how likely life is to develop on a planet that can sustain it; we don't know what the probability is that life, once originated, evolves intelligent forms (whatever, of course, "intelligence" means at the galactic level); we have no way to estimate how likely it is that intelligent life actually develops means for interstellar communication; and finally we have not the foggiest idea of how long a technological civilization actually lasts (in the latter case, we do not even have a sample size of one, since we have not witnessed the extinction of a single technological civilization—fortunately). While these quantities are certainly estimable in principle, it is hard to see how we could possibly produce reasonable figures in the near future, or how this could be achieved without technological advancements (such as, perhaps, space travel) that are not only a long way from being realized, but may themselves turn out to be unachievable.

Moreover, you might have noticed yet another unspoken assumption underlying the entire SETI project, one that again is highly questionable and for which our best answer once more is the principle of mediocrity, as unsatisfying as it may be: we have been talking so far of intelligent life, technological civilizations, and interstellar contact, all the while thinking of ET as more or less characterized by the same sort of psychology as that of human beings. But why should that be so? On what grounds do we think that intelligent beings are bound to produce civilizations, and that these in turn will develop technology, and—perhaps most crucially—will be interested in deploying this technology to send messages to potential listeners throughout the galaxy? The only reason seems to be "because we did it," which is rather weak on the face of it. In fact, if human psychology is any guide at all at the galactic level, then there are a lot of people listening out there, but nobody is broadcasting! Humanity, so far, has purposely sent only one radio message to other potential civilizations:[9] it lasted only 1,679 seconds (almost 28 minutes), and it was sent on 16 November 1974 from the largest single-dish radiotelescope in the world, in Arecibo, Puerto Rico. The message included basic information about life on Earth, humans, and the solar system, and it required a power of 1,000 kW. It was sent toward a target of questionable usefulness, the M13 star cluster in the Hercules constellation. The cluster was chosen because, as the name suggests, it is chock full of stars (several hundred thousands); on the other hand, the very high density of those stars makes it improbable that there are planetary systems in that area (too much gravitational instability). Besides, the cluster is 25,100 light years away, which is the time the message will take to get there, so don't count on being around when (and if) the answer comes.

There are various reasons why we have not transmitted repeatedly and toward a variety of targets, beginning with a dearth of political will (or even downright paranoia at revealing our position in the sky!), and therefore lack of funding. But also the obvious absence of a tangible reward: should we ever succeed in capturing an extraterrestrial signal, the payoff would be enormous regardless of what the message said: it would be confirmation that there is indeed other intelligent life out there, which would have tremendous psychological, philosophical, and even theological consequences. But what's in it for a team of scientists who *send*, rather than listen to, messages? They are certainly not going to be around if and when an answer comes back, making the whole thing much, much worse than the proverbial message in a bottle. If that sort of rather selfish, or at least self-absorbed, human psychology permeates the galaxy (again, following the principle of mediocrity), then we are not very likely to get a hello any time soon.

Time now to consider the major theoretical objection to SETI, the so-called Fermi paradox, named after the Italian physicist Enrico Fermi (better known for his role in discovering nuclear fission, and later on making the first atomic bomb). The story goes that Fermi was having lunch with some colleagues back in 1950 (notice how both the Drake equation and the Fermi paradox are very old concepts, which does not exactly mark SETI as a field burgeoning with theoretical innovation). The conversation veered to how many intelligent forms of life there might be in the galaxy, apparently with most participants taking for granted that the number should be pretty high (the Drake equation had not been published yet, but of course astronomers had already estimated the size of the galaxy at hundreds of millions of stars). Fermi listened to the arguments being made and then contributed a simple, but devastating, remark: "[Then] where is everybody?"

What Fermi meant was that if intelligent life is really common in the galaxy, we should have seen its signs by now, and the absence of any contact so far should be taken as a strong indication that we may be alone after all. This sounds like a preposterous remark, perhaps made in jest, but the argument does have some force. Fermi's point was that one ought to take seriously the vastness of time involved in the history of the galaxy. Regardless of how slowly a technological society may develop the ability to reach other stars, and irrespective of how slowly it may succeed in colonizing other habitable planets, over time scales of hundreds of millions, or even billions, of years that civilization is likely to have succeeded in reaching every corner of the Milky Way and establishing itself in every suitable stellar system. Since this, obviously, has not happened, Fermi's inference was that perhaps no such civilization exists or ever existed. Ergo, we are the only ones around.

The Fermi paradox, as this is known, has been taken seriously enough by SETI researchers to generate a cottage industry of possible explanations, none of which are particular convincing. Perhaps, some have suggested, interstellar travel is simply too costly, and conquering the galaxy in the way the Spanish conquered South America takes a lot of determination and resources. Or maybe, another explanation goes, we are in fact surrounded by ETs and simply don't know it, perhaps because they decided to make the solar system a natural reserve, a place where other races can go and see what it is like to be in the infancy of civilization (appropriately, this has been nicknamed the "zoo hypothesis"). And of course there are those who point to UFO sightings as proof that in fact we are being visited, and possibly have been for a long time. You can make up your own story, if you give it five minutes' reflection. The point of interest to us is that none of the above (including the Fermi paradox itself) smells anything like science.

There are two significant problems with Fermi's paradox, which again shouldn't require a rocket scientist to figure out (this is no reflection on Fermi's reasoning abilities, of course; after all, he was just having a casual conversation with colleagues over lunch, on a topic on which he was not doing, and never did, any research). The most obvious problem with the paradox is that it simply assumes, without justification, all too familiar human psychological attributes (particularly, a penchant for exploration and/or conquest) and projects them on other life forms of which we know absolutely nothing. This is the by now familiar principle of mediocrity, but on steroids! Perhaps the simplest explanation for Fermi's paradox is that other civilizations just don't behave like Spanish conquistadores. Maybe they have their version of the "prime directive" that has provided so much fodder for *Star Trek* episodes: don't mess with civilizations at their preinterstellar-flight stages, we'll welcome them into the galactic club when they're mature enough.

The second problem with the Fermi paradox, it seems to me, is that it is not exactly logically tight. Let us assume for the sake of argument that there is, in fact, another technological civilization out there, at about the same stage of development as ours (by the way, whoever decided that civilizations have to go through stages similar in any way to our own history?). Now imagine that somewhere on the mother planet of that civilization, a well-known physicist named Henry Stop is having a nice lunch with colleagues, and they are casually discussing the possibility of the existence of other civilizations just like theirs. Henry listens to his optimistic colleagues, who think that the galaxy teems with intelligent life, and quips: "Then where is everybody?" Henry Stop has just "demonstrated" that we on Earth do not exist. Now, that is what I call a paradox.

After all of the above, what are we to make of SETI vis-à-vis the question of how scientific the enterprise is? Well, unlike various interpretations of quantum mechanics, or string theory, SETI does allow for the collection of systematic observations that at least open up the possibility of empirical tests of the fundamental hypothesis that there are other technological civilizations out there. However, the actual likelihood of a positive test of this hypothesis seems very low, and there essentially is no way to reject it on observational (let alone, of course, experimental) grounds. Moreover, although there is some rudimentary overarching theory underlying SETI, like the Drake equation, it is far less sophisticated than other mainstream scientific theories, and it is fraught with significant conceptual and empirical problems. To make things worse, discussions about SETI are uncomfortably prone to be highjacked by "just-so" stories, hypothetical scenarios with no independent empirical validation, or even a possibility of finding such validation—as my discussion of the Fermi paradox attempted to show. SETI, though, is not alone in occupying this perilous position between science and nonscience (or even, some would argue, pseudoscience). Our next example comes from the field of evolutionary biology, but it provides many parallels with SETI. Perhaps it is because it concerns the search for the origins of another intelligence, the human variety.

Is Evolutionary Psychology a Pseudoscience?

Evolutionary psychology is the most current incarnation of what started out as sociobiology, a branch of evolutionary theory aimed at explaining the behavior of animals (and in particular humans) as the result of evolutionary processes, especially natural selection. The turning point for this research program was the publication in 1975 of *Sociobiology: The New Synthesis*, by Edward O. Wilson,[10] a first-rate student of social insects. Wilson's book would have caused little controversy if it were not for the last chapter, where he extended his approach to human beings, which he did more fully in 1978 with his *On Human Nature*.[11]

The debate that ensued was ideologically and politically biased on both sides, and it included ugly episodes such as renowned biologist Stephen Jay Gould publishing two negative reviews of Wilson's book (a behavior that borders on the unethical), and a famous incident when Wilson was treated to a shower of cold water and ice by an activist student attending a scientific meeting (so much for scientific fair play). On the other side of the divide, sociobiology's cause has not been helped by the publication of intellectually questionable works such as Richard J. Herrnstein and Charles Murray's *The Bell Curve: Intelligence and Class Structure in American Life*,[12] with its largely unsubstantiated claims about genetic determinism of human cognitive traits, and Randy Thornhill and Craig T. Palmer's *A Natural History of Rape: Biological Basis of Sexual Coercion*,[13] which proposed on the ground of very little evidence and much extrapolation that rape may be an adaptation for otherwise unsuccessful males.

From the perspective of this book, however, and as dispassionately as is possible when one considers an issue so emotionally and politically charged, the question can be asked: is evolutionary psychology a legitimate branch of evolutionary biology, or does it approximate more the status of a pseudoscience, as its critics have often contended? Let me start by saying that the overarching idea that behaviors (and therefore cognitive traits) can evolve, and sometimes do so as the result of natural selection, is completely uncontroversial among scientists, and so it should be. The issue, rather, is whether we can have sufficient evidence that natural selection has shaped any particular human behavioral pattern. The devil, as the saying goes, may be in the details. To quote one of the most preeminent critics of evolutionary psychology, geneticist Richard Lewontin, "I must say that the best lesson our readers can learn is to give up the childish notion that everything that is interesting about nature can be understood. . . . It might be interesting to know how cognition (whatever that is) arose and spread and changed, but we cannot know. Tough luck."[14]

Raising the question of whether evolutionary psychology is a pseudoscience in this context is another way to ask what characterizes pseudoscience—one of the aims of this book. John Casti, in his marvelous (and marvelously titled) *Paradigms Lost*[15] tackles the same problem and lists a series of telltale signs of pseudoscience that may be used as a reference point. These include anachronistic thinking, the glorification of mysteries, the appeal to myths, a cavalier approach to evidence, an appeal to irrefutable hypotheses, the emphasis on probably spurious similarities, explanation by scenario ("story-telling"), "literary" rather than empirically based interpretations of facts, extreme resistance to revising one's positions, a tendency to shift the burden of proof, and sympathy for a theory just because it's new or daring. Keep this list handy, because we will encounter several egregious examples of it in the next chapter.

Based on Casti's criteria, the verdict for evolutionary psychology (like the one rendered above concerning SETI) is a mixed one. Let us take as a representative example Thornhill and Palmer's idea that rape may be a naturally selected strategy for lower-ranking males to "sneak in" some reproductive fitness, considering that they would otherwise not have access to females. Regardless of how distasteful the idea is from a moral perspective, it is of course perfectly sound from a biological one. After all, there is in fact well documented evidence that "rape" (defined more neutrally as forced access to females) is present in a variety of animal species. The trouble starts when one looks for corroborating evidence for the hypothesis of natural selection in the specific case of humans. To begin with, there is actually convincing data out there to show that rape certainly doesn't pay as a reproductive strategy in today's society. This is because of abortion and because the risk to the perpetrator is very high (if caught, he can be locked away for a long time, dramatically reducing his future chances of passing genes to the next generation).

Evolutionary psychologists, however, tell us that the modern behavior did not evolve in response to the highly derived environmental conditions of today, but rather with respect to the conditions prevalent during a crucial period of human history that occurred in the Pleistocene. The irony here is that this explanation is both reasonable and essentially untestable (again, coming very close to the sort of problems we encountered with SETI). It is certainly reasonable to think that natural selection acted for a long time in premodern human populations and affected both their appearance and behavior; but unfortunately behaviors have an annoying tendency not to leave a fossil record, and neither do the details of the (largely social and cultural) environment under which natural selection allegedly operated throughout recent human evolution. This makes resorting to Pleistocene scenarios a "just-so story," with very little scientific content or relevance. (Notice that this does not imply that it is not true, just that it is not scientifically testable.)

The other major route available to evolutionary psychologists is the so-called comparative method, that is, the possibility to study the evolution of a characteristic by comparing a focal species (humans) to their close relatives (the great apes). But once again we run into the problem that, while this approach is indeed widely used in evolutionary biology, it simply happens not to work very well at all in the case of our own species. The reason is another unfortunate accident of history: for the comparative method to work properly, one needs many closely related species to compare (to achieve statistical power). Alas, we only have a handful of living close relatives (one or two species of chimps, and one or two of gorillas), and they are actually separated from us by several millions of years of independent evolution, that is, they are really not that close. It would be nice to know, but it looks like we simply don't have enough historical traces to make much progress.

The criticisms raised to the example of the rape-as-an-adaptation hypothesis are valid for many (though not necessarily all) of the other claims made by evolutionary psychologists, but this should not be taken to indicate more broadly that evolutionary biology itself is a borderline science. To make clear why, and to highlight the contrast with evolutionary psychology, let us briefly consider a successful example of an evolutionary explanation for a specific behavior related to sexual preferences. We shall see that the proposed explanation is in fact eminently testable and therefore represents a standard example of solid science. The sexual behavior in question is not human, however, but characteristic of the swordtail fish, scientifically known as *Xiphophorus helleri* and well known to all aquarium lovers.

You see, male swordtail fish have a long and colorful tail (the "sword"), which is apparently very attractive to females of the same species.[16] One can demonstrate that a long tale is favored by sexual selection, because males with artificially shorter tails are less attractive to the females, while males with longer than natural tails are even more attractive. The experiments, in other words, settle the matter of why long tails persist in this species: they are adaptive in the current social environment. Moreover, one can also do experiments showing why tails don't get even longer, despite a clear preference on the part of the females. It's because the longer the tail, the more likely the male swordtail fish is to end its life as prey to another species, in which case being very attractive to its own females would probably not do him much good. This, incidentally, is a spectacular confirmation of Darwin's prediction that sexual and natural selection will balance each other out at the point where pleasing the other sex is not too likely to turn you into lunch for someone else (think of the peacock's tail).

There is more. Once the matter of why tails are around, and why they are of a certain length rather than another, has been settled via observations and experiments in currently living swordtail fish, one can reasonably ask which evolved first: the long tail or the female preference? It would seem that both had to be in place simultaneously for the system to get started, an example of what a creationist might call "irreducible complexity." Not so: it turns out that females evolved their preference first, and that males that happened to have a slightly longer or more colorful tail were selected in favor. How do we know this, since ancient female swordfish did not leave a fossil record of their behavior? Because the swordtail belongs to a fairly large group of related species, several of which have the tail, but others of which don't. Biologists have been able to reconstruct the phylogenetic relationships (the "family tree," if you will) of these species, based on morphological and molecular data. Once they had the tree figured out, they searched for the most closely related species in which the males did not have an elongated tail. They then took the females of this species and exposed them both to males of their own species that had been augmented with an artificial tail, as well as to males of closely related species that naturally had a long tail. The results were clear: the females of species with males lacking the sword *still* preferred enhanced males with the sword. The conclusion is that the female preference evolved first, the long tail followed. (Incidentally, we do not know *why* females in this group of species have such preference: as usual in science, as soon as one question is answered, another mystery pops up. That's what makes the whole thing so much fun to begin with!)

The moral of this story is that evolutionary biologists have a variety of tools available to test hypotheses about adaptation and natural selection: in the case of the swordtail fish, one can study the fish's behavior in the wild, do experiments that manipulate the system, and even look at enough closely related species to be able to reconstruct historical events despite the lack of a relevant fossil record. What makes evolutionary psychology problematic, therefore, is not the testability in principle of the basic idea that natural selection may have shaped cognitive traits in an adaptive manner, but rather that the human species in particular is an awful model system for these studies: we cannot measure selection or do experiments in the relevant environments (because the Pleistocene was very different from any modern environment in unknown ways), we cannot compare our behavior to that of closely related species (because there are only three or four of them, and they are rather distantly related to us and to each other), and we cannot use the fossil record (because it does not tell us much about human cognition, other than very basic things, such as the size of the brain of our ancestors). It's bad luck, but one that a scientist ought to take seriously, or he risks turning a potentially scientific enterprise into bunk.

All of this does not make evolutionary psychology a typical example of pseudo-science, like astrology or parapsychology, but it certainly moves it away from mainstream evolutionary biology and into a territory uncomfortably close to purely historical research, where the production of a variety of possible scenarios is the best one can hope for, and hard data are difficult to come by or interpret. Did Napoleon lose at Waterloo because of complex military and political reasons, or because he happened to suffer from diarrhea that morning and could not concentrate on the task at hand?[17] The latter hypothesis seems ridiculous, but is in fact a possible, if not terribly plausible, scenario. The problem is that we are not at all likely to find out. As we shall see now, however, history itself, or at least some aspects of it, may turn out to be characterized by a higher degree of empirical testability of its hypotheses than evolutionary psychology, as the next example will illustrate.

History as Science?

History, some people say, is just one darn thing after another. Or is it? From time to time there are attempts at making history more "scientific" by introducing general theories of how it unfolds. Perhaps the most famous one is the Marxist view of history, according to which major historical trends and events can be understood in terms of class struggle and economic inequality. Marxists may have a point, but their theory is so flexible that Karl Popper rightly lumped it with Freudian psychoanalysis as being useless because it could explain (or, rather, flexibly accommodate) everything, and therefore, in fact, explains nothing—not different in this respect from standard pseudoscience.

A valiant attempt to build a general theory of history has been articulated by biologist Jared Diamond, first through his *Guns, Germs and Steel* (1997) and later with *Collapse: How Societies Choose to Fail or Succeed* (2005). We will briefly look at Diamond's project as he presented it in the first book, since the second one has received sustained criticism by historians for a number of alleged factual mistakes or misinterpretations, and to consider it would distract us from our central quest. Diamond's is an example of both the potential and limits of an approach that attempts to apply the methods of the historical sciences to a discipline usually seen as the least scientific of the humanities.

What does a biologist have to do with history, anyhow? As Diamond puts it, and as we have already seen, evolutionary biology (his specialty) is, in fact, a historical science, similar to astronomy and geology, but quite different from physics and chemistry. Might history itself, asked Diamond, benefit from the use of the standard methodological and analytical tools of historical sciences? Historians are arguably not the best candidates to answer that question, since they are usually not trained in the sciences at all, although of course ultimately Diamond's project will succeed only if it will be endorsed by a new generation of scientifically inclined historians. As it will become clear in the next few pages, even such success, should it ever come, will not in fact mean the elimination of history as the sort of discipline we know today, because Diamond addresses ultimate, not proximate, historical causes.

It is precisely this distinction between types of causes, familiar to evolutionary biologists and in fact introduced by Aristotle, that is at the core of Diamond's books. The guns, germs, and steel of the title of his first book on history are the proximate, that is, immediate, causes of the dominance of Western civilization in the Americas; it is through superior weapons, better metal technology, and especially the not always accidental effects of lethal diseases that the Spanish conquistadores subdued the large pre-Colombian empires of Central and South America, and that the north American British ex-colonies wiped out native Americans in the pursuit of their alleged "manifest destiny." But, asks Diamond, why did the Europeans end up with the better guns, the steel, and the more lethal diseases? Why did not the Inca emperor Atahuallpa capture King Charles I of Spain, rather than the other way around?

Diamond makes a good argument that two of the classical explanations for ultimate causality in human history are unlikely: let us call them the genetic and the latitudinal one. The genetic explanation says that, to put it bluntly, the Spanish conquered the Inca rather than the opposite for the simple reason that Europeans are intellectually superior to South Americans (and, of course, the same "argument" has been applied to Europeans versus Africans, though, curiously enough, the possibility of, say, Chinese superiority over Europeans, in terms, for instance, of historical expanse of controlled territory or precedence in several technological inventions, is rarely brought up by Western writers). Diamond finds the genetic hypothesis appallingly racist, but of course that is entirely irrelevant to its validity: if there were demonstrable and significant genetic differences in the intellectual capacities of different groups of humans it would be simply irrational (and profoundly antiscientific) to deny them. The bottom line is, however, that we have no credible evidence that such group differences exist (though of course there surely are heritable cognitive differences between individuals within groups), and the likelihood is that they are either not there or that they are rather small, and thus not likely to account for large historical trends.

The latitude hypothesis, on the other hand, belongs to a class of explanations that Diamond actually endorses, those that rely on differences in the *environment* in which different civilizations arose. The problem with the specific idea that peoples living in northern latitudes are somehow more prone to develop technologies and new ideas is, however, fraught with a variety of problems, beginning with the fact that there are plenty of exceptions (think of the more complex societies of the Inca, Maya, and Aztecs compared to their North American pre-Columbian neighbors, for instance) and that it is not at all clear why simple latitude would make a difference (is it that colder climates spur innovation because they are more challenging, or because they allow people more time to think during the long idle winters?).

Before we get to Diamond's central idea about the relevant factors affecting ultimate causality in human history, however, we have to understand his methodology, and in particular how he goes about both rejecting other explanations (such as the latitudinal one) and building a case for his own. Diamond points out that historical sciences are inherently *comparative*: since historical scientists cannot usually do experiments (one cannot rerun the events that led from an asteroid impact to the extinction of the dinosaurs), they rely on comparing multiple occurrences of similar events across large swaths of time and space. In particular, historical scientists make use of what are often called "natural experiments." Far from being an oxymoron, the phrase refers to those cases where enough similarities of circumstances occur between different events to make causal inferences possible, especially if some of those circumstances are different enough between individual instances to help the researchers rule out some causal hypotheses and support others.

In evolutionary biology, for instance, natural experiments occur in the well-known phenomenon of convergent evolution, where the same traits or behaviors evolve independently several times in response to similar circumstances. A typical example is the evolution of sleek, elongated bodies in animals faced with the problem of how to swim fast: regardless of their individual phylogenetic histories, fish, sharks, reptiles, and mammals all have evolved a similar body type under similar environmental circumstances. One can infer that this repeated pattern of evolution is likely the result of natural selection for fast swimming rather than, say, of the idiosyncrasies of each species' genetic makeup, precisely because the genetic makeups of these groups of animals are very different, and yet both their bodies and their environments are similar. Analogously, claims Diamond, if one sees the same general pattern occurring in different places and times of human history, one can examine the similarities and differences among these natural experiments to infer something about the underlying causes. The method is no guarantee of truth, but nothing in science is anyway.

The bulk of *Guns, Germs and Steel* is devoted to the deployment of a small number of explanatory principles and their application to a variety of natural historical experiments. It would be beyond the scope of this book to go into the details (besides, Diamond's book makes for excellent nontechnical reading), but we need to understand what those principles are:

1. Perhaps the most crucial principle is in fact a variation on the old latitudinal hypothesis that Diamond rejects: a major causal factor explaining the rise of more successful civilizations in Eurasia, say, than in sub-Saharan Africa or on the American continent is that Eurasia expands along an East-West axis, while the Americas and Africa geographically extend chiefly along a North-South axis. This is significant for a variety of reasons, but the main one is that plants and animals that are domesticated in one region can easily be exported and adapted to other regions at the same latitude (East-West), but often not so if they are moved along a latitudinal gradient (North-South). On top of this, Eurasia simply happens to be characterized by far fewer and less formidable natural geographical barriers than either Africa or the Americas, which again greatly

affected—according to Diamond—the ability of various populations to engage in early cultural and technological exchanges.

2. Because of biogeographical reasons, certain areas of the world (like Eurasia's fertile crescent) and not others (like most of Africa and the Americas) were characterized by a large number of wild species of plants and animals that would turn out to be easy and profitable to domesticate once people invented agriculture. For instance, both Eurasia and Africa had horselike animals: the horse itself, and the zebra. But these two species, though closely related, are different enough in temperament that it is relatively easy to domesticate the horse and next to impossible to do the same with the zebra (people have tried). Once these differences are compounded over a relatively large number of animal and plant species, one can begin to see how some civilizations took off and others got left behind.
3. Obviously, different habitable areas are of different sizes, for instance, compare Eurasia with Papua New Guinea (one of Diamond's crucial natural experiments). Larger land masses, especially if rich in natural resources, allow for much greater population numbers. More human beings means more talent and raw labor to develop and deploy new inventions, so sheer population size can make or break the difference between a successful civilization and one that does not progress past the Stone Age.
4. Finally, the proximity and accessibility of different continents from each other also makes a difference, according to Diamond: Europeans had ready access to Africa, for instance, and could mine its resources, as well as export their know-how. Australia, on the other hand, is far enough from any other continental land mass that contact remained sporadic and chancy. In the long run, repeated occasions for contact are bound to affect the history of both parties, sometimes dramatically so, while rare encounters are more likely not to be repeated and to have only local or temporary effects.

Part of Diamond's program, of course, is to connect ultimate and proximate causes, to provide for a complete, organic picture of history. The idea, again, is to draw inspiration from evolutionary biology (though this does not at all mean that Diamond wants to "reduce" history to biology). Biologists are interested in the proximate causes of biological phenomena—say, the molecular and developmental mechanisms that build a wing—but also in the ultimate causes of those same phenomena, such as the environmental circumstances and selective forces that led to the independent evolution of wings in different groups of animals, including several types of insects, reptiles, and mammals. Analogously, Diamond claims that historians should certainly retain interest in proximate causality, exemplified by the guns, germs, and steel of the title of his book. But he sees these in turn as the result of broader-scale phenomena such as the four explanatory principles I briefly summarized above.

Let us take the example of lethal germs, for one. As a biologist, Diamond knows very well that many human diseases evolve in parallel with the domestication of animals and with the development of large populations, especially when grouped in small areas (such as cities). It is therefore not by chance that the Europeans, and not the Inca or the Native Americans, were the ones who carried diseases that contributed greatly to wiping out the local conquered populations: Eurasia is where the largest numbers of domesticable plants and animals happened to be when humans began to practice agriculture and domesticate animals, which led to population expansion and to large cities, simultaneously with the evolution of disease agents to which the Europeans, but not the Inca or the Native Americans, had been exposed for generations and had developed at least partial immunity (think smallpox). That, roughly and in a simplified manner, is one of the chains of events—from broad, ultimate causes all the way down to specific, proximate ones—that explains why the emissaries of Charles I of Spain conquered the subjects of the Inca Atahuallpa rather then the reverse.

Diamond anticipated the obvious objection to his view of history, the accusation of "environmental determinism," which, in historical circles, is almost as bad as that of genetic determinism. Both fears of determinism are rooted in a persistent misunderstanding of causality in human affairs, one that again would take us too far afield to pursue in this context. But it seems to me (and to Diamond) rather strange to deny the influence of environmental (or, should they turn out to be relevant, genetic) factors in contributing to human actions in the name of a quasi-mystical concept of human "freedom of the will." Human beings certainly have the ability to make decisions and alter the course of their actions, but neither those decisions nor those actions can possibly be independent of the genetic makeup of humanity or of the environmental conditions in which human beings find themselves. To put it in another fashion, human beings are not exempt from the standard laws of causality.

One episode will, I hope, make clear what I take Diamond to mean here. A colleague of mine teaches introductory philosophy courses in which she tries to get across the consequences of inequality of resources—a small version of the same sort of causal factors Diamond is after at a much larger scale. Typically, her students are convinced that "environmental determinism" (they wouldn't use that term) is a bogus idea, that almost regardless of circumstances they will be able to succeed if only they are willing and able to work at it. Call it the American dream of self-made success against all odds, if you will. My friend then divides her class into small groups and gives them bags of materials and tools to carry out a simple project: build a mobile like those used to entertain infant children. Typically, the students launch into the project trying to come up with the most colorful, inventive, and fun mobile they can make. After a while, however, they look around and notice something odd: some groups received a full array of tools (scissors, cutting knives, pencils, tape, etc.) and of materials (colored paper of various consistencies and patterns, an assortment of magazines); other groups, however, only got a rather minimalist supply of tools, materials, or both. As soon as they realize that the playing field is not level, they begin to complain of unfairness, to which my smiling colleague replies that most of them had just finished arguing a few minutes before that it doesn't matter what the resources available to someone are, if they *really* want it, they can and will succeed. So, what's the problem in making the mobiles?

It is in this sense that one should read Diamond's argument about the ultimate causes that led to huge imbalances among civilizations in the availability of guns, germs, and steel, and that therefore have played a fundamental role in the unfolding of human history so far. You can be as smart as you want, but if you happen to live in an environment where there are few domesticable plants and animals, or where it is impossible to develop high population densities for lack of resources, or where contact with other civilizations is restricted by geographical barriers, you are much more likely to build a rather boring mobile, so to speak, compared to people living in more favorable environments.

Diamond, toward the end of his book, does two important additional things for his readers: he outlines what further work needs to be done to pursue his project, and he discusses the very nature of historical science—which brings us back to the underlying theme of this chapter, the exploration of areas of inquiry poised at the borderline between science and nonscience, some closer than others to falling on one side or the other of that line.

There are several ways in which the ideas outlined in *Guns, Germs and Steel* (and later in *Collapse*) can be pursued, expanded, and—most importantly—further tested. For instance, Diamond suggests a systematic tabulation (which he has not done) of plant and animal candidates for domestication on different continents, together with an exploration of the biological reasons they were or were not, in fact, successfully domesticated. Another avenue of research is the assembling of quantitative data on the rate of diffusion of both populations and innovations across time and space, to see if they accord with the predictions Diamond makes in his book on the basis of general principles and selected examples. These and other suggestions make the important point that the kind of general theory of history that Diamond has in mind can be tested by proper comparative analyses, just like any other theory in any other historical science.

As we have seen in the last chapter, however, one needs to understand in what sense historical sciences can make testable predictions, because it is very different from that of the physical-chemical sciences, and is so often a source of the wrong-headed conclusion that history, therefore, cannot be "scientific." Once again, Diamond resorts to an example from biology to make his point. He says, correctly, that a biologist familiar with human genetics and developmental biology can make a fairly accurate prediction of the percentage of newborn babies in any hospital at any given time that will be males: it is just above 50 percent. We know the reasons for this: it is due to a combination of basic genetics (a proximate cause—in particular Mendel's laws) and the effect of natural selection (an ultimate cause) that generates a slight deviation from the simple 50:50 ratio one would expect, because human males suffer a higher mortality than females (in turn, the difference in mortality is due to a proximate cause: males have only one copy of the X sex-determining chromosome, which means that lethal mutations on that chromosome are not masked by the second copy of the same gene that is normally carried by females). What neither Diamond nor any other biologist could do is to predict that Diamond's own two children would turn out to be males: the general principles predict (accurately) the broad pattern, but individual happenings are still governed by a good dose of chance.

Analogously in history: while the historian may be able, in certain cases, to construct and test plausible hypotheses about ultimate and proximate causes shaping broad patterns of historical events, she still will not be able to tell whether Napoleon's loss at Waterloo had anything to do with my imagined scenario relying on a simple case of diarrhea, nor will she be able to predict when and how the next supersized male ego will cause a continent-wide war. This partial ability to predict, however, does not make a scientific approach to historical research a ludicrous idea. Ironically, what may delay or even impede the sort of research Diamond is advocating is the distrust (perhaps fear?) on the part of historians themselves.

One of Diamond's critics, for instance, is Eugene Goodheart, who in his self-explanatorily titled *Darwinian Misadventures in the Humanities* devotes an entire chapter to explaining why, in his mind, Diamond's project for history is fundamentally misguided. Goodheart begins by saying, quite correctly, that "impressive as he is as an empirical historian, [Diamond] offers very little in the way of guidance to an understanding of the proximate as opposed to ultimate causes that make up the smaller, though by no means insignificant, times frames of human history." But this is quite a strange complaint, considering that the goal of Diamond's book is precisely to address the ultimate, not the proximate, causes of history!

Goodheart continues by raising what he sees—erroneously—as a fundamental objection to Diamond's attempt: for Goodheart, a crucial feature of history is that a historian cannot make predictions.[18] Hence, he concludes, history does not qualify as science. QED. But we have just seen in the last chapter that making predictions *is* possible for historical science, as long as one understands "prediction" to indicate the formulation of hypotheses that can be tested by uncovering new data, and not the much narrower conception of prediction of future events. Historical sciences, not just history, are very bad at forecasting the future, but they are perfectly capable of meeting the standard of science because (some of) their hypotheses are eminently testable.

Goodheart further mistakes the nature of science as being based on the formulation of "natural laws"[19] and as being "mathematical,"[20] claiming that since history does neither, then historical research cannot be done scientifically. These are common misconceptions about science, often propagated by scientists themselves. But in reality there is plenty of good science that is done without mathematical formulation (because mathematics is a convenient language to formalize certain problems, but not a necessary condition for doing science) or that does not lead to the formulation of any universal law (indeed, physicists themselves are beginning to question whether the so-called laws of the universe are truly universal or instead apply locally, in either time or space).

The general point is that historical research may not develop along the lines suggested by Diamond not because it cannot be done, but because historians themselves—at least of the current generation—seem to be hostile to the very idea of testing their hypotheses on empirical grounds. As we shall see later, this attitude of mistrust of science from within academia itself is a major cause of blunders in public discourse and has brought about the so-called science wars of the last few decades.

More recently, the idea that history can and ought to be turned into a more rigorous, scientific discipline has been argued by another scientist: Peter Turchin, a theoretical ecologist at the University of Connecticut.[21] Turchin complains that there are more than two hundred explanations proposed for the collapse of the Roman empire, a situation he finds "as risible as if, in physics, phlogiston theory and thermodynamics coexisted on equal terms." He even proposes a new term for his approach, to distinguish it from the type of history that is done in humanities departments: "cliodynamics," from Clio, the muse of history for the Greek and Romans. The "dynamics" part refers to the central concept advanced by Turchin that history is characterized by regular and predictable patterns, from which we can learn and that we can predict.

With two of his colleagues, Turchin has carried out some quantitative work of the type that Diamond did not get to, and he has in fact found a statistically significant trend across various societies according to which "the number of instability events [wars, eruptions of violence] per decade is always several times higher when the population was declining than when it was increasing." This result was obtained by studying societies and time periods as different as the Roman Empire and eight Chinese dynasties. This is a promising beginning indeed, and yet Turchin's own example shows how difficult this sort of research really is: his work has uncovered an interesting pattern, but has not come even close to testing specific causal explanations for that pattern. Pattern discovery is of course crucial to science, but Turchin himself admits that "the connection between population dynamics and instability is indirect, mediated by the long-term effects of population growth on social structures," going on to list a good number of interacting causes that may underlie the pattern he uncovered. The question is how is one going to test a multicausal, multilevel hypothesis with the chronic paucity of comparable data that characterizes history? In this respect, the situation is similar to the discussions among paleontologists about what causes, say, repeated mass extinctions, where people disagree on the very number of ascertained events of this kind during Earth's history, and where the number of occurrences is barely sufficient to reach statistical significance as a pattern, let alone to provide enough discriminatory power among competing causal hypotheses. Nonetheless, the proof is in the pudding, so to speak, so I wish Turchin and his cliodynamics much luck, and we shall all follow with interest whatever else he, Diamond, and colleagues may turn up over the next few years.

To return to the broad problem of the nature of science, we have seen that there is a complex part of the intellectual landscape that occupies a transitional zone between science proper and activities that may not be entirely "scientific" by the criteria we have laid out so far (systematic observation, formulation of general theories, empirical testability of specific hypotheses). Some of these quasi-sciences are nested within canonical scientific disciplines, as in the case of evolutionary psychology within evolutionary biology, or the SETI program within astronomy and biology, or even more obviously in the example of string theory within theoretical physics. In the next chapter, however, we will decidedly leave the area of science proper to briefly explore the fascinating realm of truc bunk.

CHAPTER THREE

Pseudoscience

But the fact that some geniuses were laughed at does not imply that all who are laughed at are geniuses. They laughed at Columbus, they laughed at Fulton, they laughed at the Wright brothers. But they also laughed at Bozo the Clown.—Carl Sagan

It has a 100 percent cure rate for . . . —A pseudoscientist

"Here's a forecast for a particularly bizarre consequence of climate change: more executions of witches." This chilling opening is from a *New York Times* column by Nicholas Kristof.[1] The article cited a study by an economist at the University of California, Berkeley, Edward Miguel, who found that women are routinely murdered as witches in rural Tanzania, but that the rate of killings increases when there is a drought or a flood, that is, whenever the living conditions for the local populations are dire and depend on the quirks of weather.[2] Kristof goes on to suggest that there is good reason to think that historical European witch hunts were also triggered by environmental factors, coinciding with crop failures. Indeed, according to the same column, research conducted by Emily Oster, an economist at the University of Chicago, shows a strong correlation between cold dips in climate in Europe and witch hunting in the period between 1520 and 1770.

Despite widespread skepticism about global warming, it took little more than a month for Kristof's prediction to come true: an article in the

Italian newspaper *la Repubblica* published on 21 May 2008 reported on fifteen women accused of being witches burnt at the stakes. Only the location was slightly off: Kenya instead of Tanzania. According to the original report by the Agence France Presse, a mob of about a hundred people went "blind with rage" in a village three hundred kilometers west of Nairobi, picked up the women in their homes, tied them up, beat them, and finally burned them alive.

The external trigger in these cases may very well be an environmental event, such as a flood or a period of drought. But those women were killed because superstition kills. Ever since I became interested in pseudoscience (which can be pinpointed to coincide with my move to Tennessee in 1996, near the site of the antievolution "monkey trial" of 1925), I have often encountered a casual and pernicious attitude toward pseudoscience and superstitious belief in general. Even educated and progressive people have a tendency to think of "skeptics" as asocial curmudgeons bent on denying any positive new knowledge unless it comes through the "orthodox" channels of anointed science. I was once invited by a talk-show host on a very popular local radio station in Knoxville, Tennessee, together with *Skeptic* magazine publisher and author Michael Shermer. We chatted amiably with the host for a while before the beginning of the show, as is common practice in these cases. As soon as we were on the air, however, the host felt compelled to share a surprising discovery with his listeners: skeptics can laugh and be generally pleasant! While most of the broadcast was about the ever popular (in the Bible Belt) creationism versus evolution "controversy," the host did ask us a question that I have heard countless times since: so what if people believe in pseudoscience or are superstitious? Who are they hurting anyway? Ask the fifteen women burned to death in Kenya, to begin with.

While not often lethal, faulty thinking about how the world works can hurt plenty. People can be swindled out of significant sums of money by practitioners of "alternative" medicine (alternative to what? to practices based on evidence?), can make really bad decisions about their lives if they listen to an astrologer, and can be taken advantage of emotionally and financially by "psychics" who claim that they will put them in touch with loved ones who have died. Everyone has a right to be irrational, but rampant irrationality in a society can be highly wasteful and destructive, and giving a pass to credulity on the grounds that "it doesn't hurt anyone" is, well, not a very rational position to take. In this chapter we will briefly examine various aspects of pseudoscience, from the new AIDS denialism to more canonical beliefs in astrology, parapsychology, and UFOs. This is of course not meant to be a comprehensive treatment of pseudoscience—to which many excellent books have already been devoted.[3] The emphasis, instead, will be on what these beliefs have in common and how they differ from the sciences and quasi-sciences we have encountered in the previous two chapters. All, of course, while attempting to maintain that relaxed perspective that so confounded my radio host in Knoxville.

Death by Pseudoscience: AIDS Denialism in Africa

President Yahya Jammeh, who has ruled Gambia for more than a decade, has suddenly found a cure for AIDS. Indeed, he has opened shop to treat AIDS on Thursdays and asthma on Saturdays, the rest of the week being devoted to political matters. The cure, according to Emma Hurd of Sky News, consists of "a rub down with [a] cream, a splash on the face with another potion and a drink of a murky looking liquid."[4] The remedy is, not surprisingly, endorsed by Gambia's Department of Health and we are assured by Jammeh himself that it works. "One hundred per cent the President can cure everyone. It is absolutely medically proven," said Minister Tamsir Mbowe. Apparently, though, independent verification is not welcome. Jammeh told Hurd that "I do not have to convince anybody. I can cure AIDS and I will not explain it to those who don't want to understand." But plenty of people do want to understand, Mr. Jammeh, and, yes, you do owe the world and especially your patients ("victims" would be a more appropriate term) an explanation of how the alleged cure works. Failing that, you should be prosecuted as an international criminal.

But President Jammeh is far from being a loner in the increasingly dangerous AIDS denialist movement. On the same date as the previous report, another Sky News journalist, Ian Woods, wrote about a miracle cure being offered by the Orthodox Church of Ethiopia. In this case it is "holy" water (i.e., water), being administered to thousands, some of whom are brought in chains by relatives because they are skeptical of the proffered cure. The treatment consists of priests hurling water to the faithful (and the recalcitrant), while at the same time beating them with wooden crosses, for good measure. Interestingly, while men must be totally naked, women can wear panties, though the medical principle behind this gender-based discrimination is not entirely clear. The priests are also—unconscionably—telling people that they can't use standard "Western" medications because they interfere with the action of the holy water, something disputed by Dr. Amone Wodoson at one of Addis Ababa's major hospitals, who—trying to strike a compromise while saving people's lives—went on record to say that "there is no adverse interaction between the two" (of course not, water doesn't adversely interact with anything in the human body, unless one is about to drown). Not everyone can be cured in this way, however, since women with wigs are "obviously" possessed with demons, and the water wouldn't work on them. Menstruating women and people who have had sex recently are also turned down. If they are lucky, they can make it to Gambia by the following Thursday, to be cured by President Jammeh.

Besides Ethiopia and Gambia, however, the big story as far as AIDS denialism is concerned has been South Africa. As a complex nation with a tense history of racial relations, to say the least, perhaps it was inevitable for South Africa to become fertile ground for a rejection of Western medicine in favor of local traditions and solutions. Still, it is simply astounding to discover the depths of irrationality reached by some South African leaders—and the absurd cost in human lives that their inane policies are directly causing (once again, it would seem appropriate to invoke a United Nations condemnation for crimes against humanity, but I'm not holding my breath). Michael Specter of the *New Yorker* published an investigative report[5] so frightening that I can hardly do it justice here. It begins with a truck driver's "vision" (a dream), in which he was instructed by his grandfather to put together a concoction to cure AIDS. The truck driver, Zeblon Gwala, then set up shop in the city of Durban, posted a "Dr. Gwala" sign on the door (despite not actually having a medical degree), and his "HIV and AIDS Clinic" opened for business, attracting hundreds of people every day and equally certainly condemning them to death by their fateful choice of magic over science.

How is this possible in an advanced and economically thriving country like South Africa? Because of the positions taken by former President Thabo Mbeki and by his then (until September 2008) health minister, Manto Tshabalala-Msimang, among others. Their attitude has been that antiretroviral drugs, which have been medically tested and shown to be effective against HIV, are poisons deliberately marketed by Western pharmaceutical companies. Moreover, according to the pair—and contrary to almost the entire medical-research profession—there is no evidence that HIV causes AIDS, which instead is just another lie spread by Big Pharma (with the help of the CIA, naturally) to sell their products. Mbeki and

Tshabalala-Msimang insist that salvation can be found in local knowledge such as the remedy that came in a dream to "Dr." Gwala. This while 5.5 million people—out of a total population of 48 million—are infected by HIV in South Africa, a huge humanitarian disaster unfolding in slow motion under our (and Mbeki's) eyes.

Of course, Mbeki's and Tshabalala-Msimang's absurd notions do have some support from a minority of academics (in a similar vein, we will see later on in the book that one can always find critics of global warming or evolution with legitimate academic credentials, if one looks hard enough). Science is a human activity, and human beings can legitimately hold different opinions about empirical evidence. Of course, sometimes the dissenting opinion is motivated by a thirst for fame, financial gain, or sheer obtuseness. In the case of AIDS denialism, the biggest academic dissenter is Peter Duesberg, a molecular biologist at the University of California, Berkeley, and the discoverer of the fact that some retroviruses (the same kind of virus that causes AIDS) can trigger the onset of certain types of cancer. Duesberg expressed his skepticism on the HIV-AIDS causal link back in 1987. This was only three years after the first published claim in favor of a connection, and Duesberg's paper at the time was a legitimate dissenting opinion published in a respected academic journal, *Perspectives in Cancer Research*. The problem is that Duesberg is stuck on his 1987 position, disregarding the overwhelming evidence put forth by literally thousands of studies published since. It is hard to know why Duesberg holds to his initial skepticism, whether out of simple stubbornness or because of the modicum of fame that such position has brought him or for the sheer pleasure of playing heretic. What is important is that his position is giving ammunition to inept leaders like Mbeki and indirectly killing millions of people.

Clearly, the story here is enormously complicated by intricate psychological and sociological factors. Again, it is hardly surprising that people emerging from an apartheid regime may be inclined to suspicion of white knights in shining armor coming to their rescue, and may wish instead to emphasize their own traditions and practices. Big Pharma is also far from spotless, and the practices of international pharmaceutical companies have been under fire for years even in the West. The search for profit at all costs often translates into literally inventing new medical "conditions" out of thin air or aggressively marketing "new" drugs that are actually trivial variations of existing ones. Increasing reports of undue pressure exercised by the pharmaceutical industry on scientific researchers, which in several cases has culminated into halting by means of legal threat the publication of data showing that a new drug was in fact harmful to an unacceptable degree, have tarnished the image of the entire sector.[6]

Nonetheless, one cannot help but find it ironic to the utmost degree that people like Mbeki and Tshabalala-Msimang resort to quoting the very white and very Western Peter Duesberg in support of their claim that there is a vast white Western conspiracy against African peoples. This is a clear example of why fighting pseudoscience entails more than just science education or critical thinking. People believe in shamanism, UFOs, telepathy, astrology, and creationism for reasons that go well beyond their poor grasp of science and difficulty in deploying the tools of critical thinking (though certainly more appreciation of both science and logic would not harm our planet even one bit). But the AIDS denialism case is also a terrifyingly clear example of why superstition is dangerous to the point of lethality, and hence something much more important than just the obsession of a small number of self-professed skeptics who think they know better than everyone else.

An additional irony of the debate about HIV-AIDS is that HIV is a spectacular example of evolution happening right now and dramatically affecting our lives, despite the millions of people denying that evolution itself takes place. HIV—if you believe scientists who work on the problem—has "jumped" from monkeys to humans only in the middle of the twentieth century, and has managed to kill already more than 25 million people, infecting at least twice as many. The reason the disease is so difficult to cure is because the virus evolves very fast, and so it continuously adapts to whatever barriers the natural response of the human body and modern medicine throw at it. So far the most effective treatment comes with so-called cocktails made of several drugs attacking the virus simultaneously. The rationale behind the cocktail idea is exquisitely evolutionary: even a rapidly evolving organism can only cope with so many environmental challenges (drugs, in this case) at the same time, and the more problems it has to solve the less likely it is that random mutations and recombination—the two mechanisms producing genetic variation in all organisms—will be able to yield an effective solution to the new environmental threat.

Even more interesting from a scientific perspective, there is now good evidence to think that retroviruses in general have played major roles in the evolution of mammals (again, if you believe in evolution over millions of years).[7] Most mammals have placentas to protect their developing embryos, and placentas evolved from modified eggs. Thanks to placentas, embryos started to develop literally as parasites inside the mother's body, protected not only from outside threats like bacteria and viruses, but also from the mothers' own immune systems. In the 1970s scientists discovered retroviruses on the placenta of various species from humans to mice. At first the discovery was puzzling and largely interpreted as yet another example of how pervasive the viral threat is to mammals. But a much more intimate connection was subsequently discovered at the molecular level: retroviruses in the placenta are found in the syncytium, a layer that constitutes the major barrier between mother and fetus. The protein that allows the fundamental property of cell fusion that makes placentas possible in the first place is called syncytin, and also provides the molecular mechanism that allows retroviruses to bind to the cells they attack. Some biologists are now suggesting that retroviruses played a major role in the evolution of the placenta itself, which means that one of the biggest threats to our health and survival may have also been a catalyst for crucial steps in mammalian evolution. It is a splendid example of how science can make sense of the complexity, beauty, and even irony of life. How much more convincing than a charlatan claiming to cure AIDS on Thursdays by using a murky liquid of suspicious origins.

Not in the Stars: Astrology as Bunk

Next on our brief tour of pseudoscience is a venerable one: astrology.[8] In some senses, astrology is almost too easy a target for the skeptic, since, as we will see in a moment, it is both hopelessly flawed theoretically and it demonstrably does not work in practice. Still, one out of ten thousand people in Western countries take it very seriously,[9] wasting countless hours of their active lives while being swindled out of millions of dollars to pursue it. Moreover, it is an almost perfect example of pseudoscience, because its claims can in fact be tested, have been tested repeatedly and shown to be wrong, and yet people continue the practice.

Let us start with a brief examination of the principles behind astrology. Historically, the idea dates back at least to the Chaldeans and Assyrians of three millennia ago, although the modern codification of twelve zodiacal signs was done by the Babylonians around 450 BCE. Still, astrology was not well developed or particularly popular until it was spread in the Hellenistic world by Alexander the Great, eventually making it into ancient Rome. Simply based on the fact that the basic principles of astrology were laid down in a prescientific era, one should be skeptical of how well such ideas might stand scrutiny in the twenty-first century, when humanity can count on sophisticated theories and observational instruments in both physics and astronomy (the "wisdom of the ancients" is often overestimated). But let us proceed one step at a time.

To begin with, it should be noted that star constellations, which are a centerpiece of the entire astrology edifice, actually do not exist. What I mean is that the constellations identified by the ancient Babylonians are optical illusions, because the stars that make up any given constellation are positioned at very different distances from Earth (though they are all within our galaxy), and the only reason we group them together is because of a projection effect. This is analogous to, say, observing the moon over Manhattan and concluding that the moon is *right over* (as in, at a short distance from) the Empire State Building. This is obviously absurd, but the claim that a constellation is an actual ensemble of stars is by no means less silly. Moreover, the demarcation lines between constellations are themselves completely arbitrary. It is instructive to note that different cultures "discovered" completely different constellations in the sky, a fact that is more consistent with the idea that constellations are a whimsical projection of the human mind than a reflection of astronomic reality. Additionally, today's astrologers can, if they wish to, purchase a small telescope or a pair of binoculars and check for themselves that there are many more stars in the sky than were known to the Babylonians. Are these "new" stars (they've been there for a while, obviously) part of the constellations? Should they not thereby affect astrological calculations?

Even if one understands the constellations to be meaningful, astrologers seem to ignore yet another fatal flow in their reasoning: the zodiac has shifted position during the last 2,500 years because of a well-known astronomical phenomenon called the precession of the equinoxes. The earth, like any other planet, has an axis of rotation (which causes the alternation of day and night) and an axis of revolution (around the sun). These two axes are not parallel, but diverge by a little more than 23°, a discrepancy that has several consequences, including the fact that we have seasons. An additional consequence is that the positions of the zodiacal constellation shift imperceptibly over centuries. The shift accumulated since the Babylonians is such that the position once occupied by a given constellation at a particular time of the year is now occupied by the *previous* constellation in the zodiac! This means, for instance, that my astrological sign should not be Capricorn (I was born on the 16th of January), but Sagittarius. Well, then, which sign should I look up when I open my Sunday paper, I wonder?

These are strong, in fact definitive, theoretical arguments, but they do not seem to convince either the astrologers or their devotees. "But look at the huge effects that the moon has through the tides," they will say, "is that not enough evidence to believe in cosmic influences on earthly affairs?" Well, it depends on what one means by "cosmic influences" and what kind of "affairs" we are talking about. The moon argument is an example of common sense applied to a problem that requires more than just common sense, but it is precisely these sorts of arguments that are often deployed by people who accept pseudoscientific claims, and the reason is that such arguments are indeed superficially very convincing. Similarly, defenders of creationism will tell you that it is obvious that living organisms are the result of the actions of an intelligent designer. In fact, living organisms *look* like well-designed organic machines adapted to their environment. Global warming deniers will make a similar argument and ask you why, if the earth is warming, do we see harsher winters in some areas of the planet. Again, a very reasonable question. The answer to evolution deniers is that natural selection results in organisms that are well adapted to their environment without the need of a conscious designer; and global warming skeptics should realize that harsh winters in some areas of the globe are precisely what one would expect if the planet is warming up (because the earth's climate doesn't work like a simple human appliance, but is instead characterized by complex—but statistically predictable—nonlinear effects). The problem is that these scientific answers are counterintuitive, like much of science itself.

Let us analyze the "moon defense" a bit more closely then. Tides are indeed caused by the gravitational effects of the moon and the sun. Although the sun is much bigger than the moon, and thus has a much stronger gravitational field, it is also much further than the moon from Earth. On the balance, the tidal force exerted by the sun is about 46 percent of that exerted by the moon. Of course, the other planets of the solar system also exert a tidal influence on Earth through their own gravitational fields. But since the other planets are both distant and small (compared to the sun), their gravitational effects are essentially negligible.[10] Tides exert bona fide cosmic influences on human affairs, as anyone who lives in coastal areas will readily tell you. But nobody has ever demonstrated a direct gravitational effect of the moon-sun system on the behavior of human beings (werewolves don't count), and simple gravitation is certainly not what astrologers are talking about when they claim astral influences, otherwise the moon would be twice as important as the sun in astrological charts, and the planets wouldn't figure at all.

What then? As astrology critic Phil Plait points out, physicists recognize three more forces in nature (other than gravity): the strong and weak nuclear interactions, and electromagnetism. That's it—there are no other known physical forces to play with. We can exclude the strong and weak nuclear forces, because they can only be effective at the scale of atomic or subatomic, not astronomical, distances. Electromagnetism is a more promising candidate, but only at first glance. Again, this is something that science has studied for a long time, and we can measure the electromagnetic fields of the various planets and of the sun. Turns out that the latter is the only one that has any significant effect, as can be seen every time that there is a solar magnetic storm, a phenomenon capable of interfering with our communications and in some cases even causing blackouts in major cities. But if electromagnetism is the conduit of the astral influences then astrology should be based just on the sun—forget the moon, planets, and other stars. Clearly, that is not compatible with astrological practice either.

It is always possible at this point for the astrologer to retreat one step further and claim that astral influences are mediated by a fifth, yet undiscovered force. We cannot be certain that there is no such thing as a fifth force, but everything that has happened in theoretical and experimental physics over the past century indicates that there are only four fundamental forces, and in fact, if anything, physicists are convinced that the four forces are really different aspects of *one* fundamental force, which they are hoping to characterize mathematically very soon (in the so-called theory of everything). If astrologers really take the bold step of postulating a fifth force in direct contradiction of modern physics, they are making an extraordinary statement backed by extraordinarily little evidence. Moreover, even if there were a fifth force, it would appear logical to assume that—*like every force known so far*—its action too depends on distance, meaning that the further away two objects are the less strength force X will have. If this is the case, then astrologers would still not be out of the woods, because they insist on treating all planets as having the same influence on human affairs, regardless of distance (could this be a leftover from a time when our ancestors thought that all stars were at the same distance from us, fixed on the same celestial sphere?). This means that not only do they have to assume the existence of a fifth force, which is improbable enough, but also that this would be the only force whose action does not decrease with distance, thereby piling improbability upon improbability. In fact, come to think of it, recent discoveries in astronomy make even this last desperate way out not an option: as I mentioned while discussing SETI, scientists have discovered upwards of three hundred *new planets* orbiting around other stars in the galaxy, not to mention tens of thousands of asteroids within our own solar system.[11] If the effect of force X does not depend on distance, then *all* these additional planets should be taken into consideration when preparing an astrological chart. Have you ever seen any astrologer drawing charts so complicated? Neither have I.

Theory in science, however, only goes so far. Philosophers have long abandoned the idea that knowledge about the world can be gathered by just thinking about it, and have yielded to the more complex approach of modern science, where empirical evidence and theoretical advancement continually interplay with each other. One could therefore argue that while astrology does not have a good theory to back it up (indeed, no sensible theory at all), it might still *work* in practice, in which case it is up to scientists to take it seriously, investigate it, and then try to come up with a theoretical explanation for why it works. Contrary to what many people might believe, scientists have in fact taken the claims of astrology seriously enough to conduct a variety of studies, close to a hundred of them. The results are out, and it's not good news for the astrologers.

Researchers Geoffrey Dean and Ivan W. Kelly have looked at the available literature and cumulated the results of dozens upon dozens of studies of astrological claims conducted in a variety of settings.[12] For example, a set of twenty-five studies involving five hundred astrologers examined the average degree of agreement between astrological predictions. In social science, such as in psychology, tests that have less than 0.8 (i.e., 80 percent) agreement level are considered unreliable. Astrology's reliability is an embarrassingly low 0.1, with a variability around the mean of 0.06 standard deviations. This means that there is, on average, *no* agreement at all among the predictions made by different astrologers. As if that were not bad enough, the same authors compared another set of forty studies on nearly seven hundred astrologers, who had worked on 1,150 birth charts. When the birth charts were matched with personality profiles, the degree of accuracy was 0.05, with a variability of 0.12—again, not statistically different from zero.

But, you might say, when *I* was given a chart by an astrologer it was so remarkably accurate! Dean and Kelly looked into that as well: ten published studies were based on the idea of allowing test subjects to pick the chart they thought matched them best, without knowing for which sign the chart was actually prepared. The "accuracy" of the pickings was a mortifying 0.002 (if you are curious, the standard deviation was 0.038, which means that the spread around the mean was even bigger than the alleged effect itself).

Dean and Kelly also examined the results of what would appear to be the ideal test for astrological predictions: they considered studies of so-called time twins, that is, individuals who are born five minutes or less apart. These people should have essentially identical astrological profiles, although it is not at all clear why *birth* is so important to begin with: do the stars and planets have no effect in the womb? What about after the individual is born and is developing into an adult? At any rate, again the results were not good news for astrologers: despite examining a whopping 110 variables in a sample of 2,011 time twins, the authors found no detectable effect at all.

Another ingenious scientific study of astrology was carried out by David Voas in England and Wales, exploiting the fact that the local census reports all sorts of data about individuals, including their birth dates and marriage statuses.[13] Voas reckoned that if the stars have anything to do with it, one should find a nonrandom distribution of the signs of married couples in the sample. In other words, one would expect some sign-to-sign marriage, say Capricorn to Aries, to occur more (or less) than by the chance rate of one out of twelve times. The sample of married couples examined by Voas was very large by any statistical criterion: more than twenty million individuals. Once Voas corrected for known biases in the sample (like the peculiar habit of assigning a birthday of January 1st to someone if their actual birthday is not known), the results are—surprise, surprise!—that there is not even the slightest tendency for people to marry according to astrological signage.

One more, just to round it up nicely. This is perhaps the most damning study of them all, because it was conducted under the most rigorous scientific standard, the double-blind experimental protocol, and was published in the 5 December 1985 issue of *Nature*, arguably the premier scientific magazine in the world. The author was Shawn Carlson of the University of California, Berkeley.[14] A double-blind test is a situation in which neither the experimenter nor the subject knows who is receiving the experimental treatment and who the control. This virtually eliminates the possibility of bias: the subject cannot guess in the "right" direction when answering a question, and the experimenter cannot interpret the data in a way favorable to one hypothesis or the other before all analyses are completed. For his study, Carlson selected thirty of the best-known and highly respected (by their peers) astrologers in Europe and the United States. Each astrologer was given three charts, one matching a target subject, the other two chosen randomly. The idea was to see how often these top astrologers would pick the right chart for their 116 clients, whom they could not, however, see or interact with. The result was once again very clear and very unfavorable to astrology: the "experts" picked correctly only once in three times, exactly as you would do by randomly choosing one of the three charts on offer. I do not know this for a fact, but I doubt that said astrologers looked at the published results and retired from their "profession" in shame, as they should have done.

Astrology can best be thought of as a psychological practice based on an astronomical theory. The theory from real astronomy shows that astrology is fatally flawed, and the empirical evidence equally clearly establishes that it does not work as a practice to help human beings. Yet, as everyone knows, astrologers keep making money off their trade, and millions of people keep wasting their savings. Why? The reasons are various and complex. One is the ability of astrologers—like psychics—to perform "cold reading," that is, of guessing a few relevant facts from the demeanor or the talk of their client and extrapolating from there. We also know well, because of controlled psychological studies, that many people have a tendency to remember positive hits by the astrologer or psychic, and yet quickly forget the much more numerous misses: this rapidly builds an impressive, if illusory, record for the astrologer. But at the bottom, astrology, like much other pseudoscience, is a largely subconscious ego trip: we feel good about the idea that stars and planets have a direct effect on our lives, and perhaps we enjoy a sense of revenge of the everyday person at telling scientists that they don't, after all, know everything under the sun (the latter part is, of course, true). And speaking of cosmic ego trips, our next example of pseudoscience also has a cosmic dimension, but with the added twist of alien intelligence being (allegedly) involved.

Death from Outer Space, Sort Of

Thirty-nine people committed ritualistic suicide as a group on 26 March 1997, in Rancho Santa Fe, an upscale community in San Diego, California.

They died thanks to a combination of vodka and phenobarbital, and they used plastic bags over their heads to cause asphyxiation just in case. They were all dressed alike (black shirts, sweatpants, and black Nike shoes), and they died in a highly organized sequence, so that later members would help cleanup the early ones (except, of course, for the very last ones in the group). They were members of the Heaven's Gate UFO cult headed by Marshall Applewhite (who was one of the thirty-nine) and Bonnie Nettles (who had died in 1985). Two more former members of the group committed suicide later on to follow in the footsteps of their companions.

Suicide, as the French writer-philosopher Albert Camus once famously said, is the only serious philosophical problem, meaning that philosophers have to account for why—given the lack of meaning in life in Camus' view—we don't all do it. The answer comes, of course, from a simple combination of common sense (what's the hurry? We will all have to get there anyway) and biology (a strong instinct of self-preservation, which we share with all life forms). The men and women who died in Rancho Santa Fe that day, aged twenty-one to seventy-two, had apparently found a different answer than most of us to Camus' question, though their reasoning would surely have surprised the French writer. Heaven's Gate members left the earth for a different plane of existence, as they put it, because they were convinced that a UFO was tracking the comet Hale-Bopp (then passing near our planet) and that the UFO would pick them up to take them to said plane. Alan Hale, an astronomer who co-discovered the comet, said that the alleged UFO was actually a star, not an intelligently guided object "tracking" the comet, as had been falsely stated on the Art Bell radio show, whose host was notoriously sympathetic to a variety of nonsensical notions.

Applewhite and his followers acted in a very deliberate way. Their group was run as a monastery, with severe rules of abstinence (seven of the members were castrated, with Applewhite providing the example). The organization was based on strict authoritarian rules, an uncompromising code of behavior, and of course active discouragement of critical thinking.[15] Applewhite had somehow managed to convince the others that he and Nettles were actually aliens incarnated in human bodies (though it wasn't clear why Nettles had "left" more than a decade earlier, without the help of a comet). In an interesting connection with millenarian Christian beliefs, Heaven's Gate doctrine included the idea that Jesus too was an extraterrestrial, and that his mission, like that of Applewhite himself, was to help humans reach the true kingdom of God.

According to news reports, members of the sect believed in an assortment of pseudoreligious and paranormal ideas, including resurrection (obviously), astrology, and psychic powers. They were also avid watchers of paranormal shows on TV. This of course does not imply that watching *The X-Files* leads to suicide, just like millions of people playing *Grand Theft Auto* video games are not automatically turned into criminals on a rampage. Rather, one suspects a number of underlying causes that contributed to the tragedy, including an abysmal (and fatal, in this case) lack of critical thinking and a desperate desire for something better than what the world was already offering to them. Videotapes left behind by group members contained phrases like "We are looking forward to this. We are happy and excited" and "I think everyone in this class wants something more than this human world has to offer." It is this strong emotional need, coupled with the inability to deploy logic or simple common sense, that is at the root of this and similarly sad episodes in the history of pseudoscience. It turns out that even UFOs can kill, not with blasting lasers a la *War of the Worlds*, but simply through the credulity of otherwise normal and intelligent human beings.

Of course, belief in UFOs does not usually turn people into suicide victims, but the same twin elements of lowered critical thinking and heightened emotional dissatisfaction with the world as it is play a role in the cult at large. There are so many alleged cases of UFO sightings, "close encounters of the third kind" (i.e., situations in which actual aliens are allegedly observed), and even abductions of humans by the ETs, that books upon books have been written to both make the case in favor of ufology and to debunk it.[16] We will take a brief look at three such cases, not only because they are representative of this vast literature, but more importantly because they will teach us some generalities about ufology as a pseudoscience. Before we start I have to make clear that a typical defense of ufologists is that no matter how hard the skeptics try, they have not succeeded in explaining *all* the alleged UFO and alien sightings. Of course not. This is precisely the same logical fallacy that we will see underlie the Intelligent Design creationist movement: the idea is to shift the burden of proof from the person who makes the extraordinary claim (to whom such burden logically belongs) to the person who simply asks for the evidence before accepting the belief. This error is pernicious because it is very common not only in pseudoscience, but in political and everyday discourse. Indeed, if I had to pick one logical fallacy I wish we could magically erase from the repertoire of humankind, this would be the one. But alas, I don't believe in magic, so I have to content myself with writing books.

The first case we will consider is one of those spectacular apparitions with thousands of witnesses, a case that James Oberg said involved "probably the largest audience of any UFO ever witnessed in North America."[17] The reason for so many witnesses is that it was a clear evening during which many people in the eastern half of the United States were outdoors to enjoy the Perseid meteor shower that had been announced by astronomers. It was around 10 p.m. on 12 August 1986. People from Georgia to Texas and from Kentucky to Quebec saw the cloudlike UFO moving from right to left in the night sky, and in Kentucky someone reported to the police of having heard an explosion during the event. The witnesses' descriptions were not always congruent, as it is often the case, but clearly many people did see *something*, because there were too many witnesses and their reports were sufficiently coherent even when filed hundreds or thousands of miles apart. Of course, it is a big step from acknowledging that an unidentified flying object (literally, a UFO) had surely been observed to claiming that an extraterrestrial intelligent machine had spectacularly shown itself that night.

What really happened became clear very soon, precisely because so many people had seen the strange phenomenon and the investigation started immediately: the UFO was actually a new Japanese rocket that had just been launched and that, at an altitude of about a thousand miles at the time of the sightings over North America, was still sunlit. The first positive identification was made possible by amateur radio listeners who received the satellite's signals. Eventually, the North American Aerospace Command (NORAD) released precise tracking data that positively identified the UFO as the Japanese satellite. But wait, what about the explosion in Kentucky? A coincidence, as it turned out: the local sheriff investigated and found that the explosion had been caused by illegal fireworks, not an uncommon occurrence in that part of the country.

There are three major lessons to be learned from the great 1986 UFO sighting. First, thousands of people are unlikely to all be lying at the same time, which means that something real happened that night; but equally obviously, thousands of inexperienced observers of the sky can be wrong about what they are actually seeing. This by itself goes a long way toward explaining most UFO cases. Second, the Kentucky explosion shows how easy it is to causally associate two events just because they take place in rapid succession, even though there is no actual connection between them. This is a well-documented logical fallacy to which we are all prone, the *post hoc ergo propter hoc* (after this, therefore *because* of this) invalid conclusion.

Lastly, as I mentioned above, one of the reasons the case was solved at all is because it was investigated immediately, when the various testimonies were vivid and when one could carry out some in-depth fact checking that logically triangulated to the right solution for the case.

Not all UFO mysteries can be solved immediately, however. Just like in the best tradition of criminal investigations, there are also "cold cases" that, with a bit of luck, acumen, and some patience, may turn alleged aliens into something much more mundane. On 12 September 1952, Mrs. Kathleen May of Flatwoods, West Virginia, together with six children aged ten to seventeen (including her two sons), went up a hill where the teens had seen a light in the sky disappear while playing football. Here is how investigator Joe Nickell summarizes the scene that presented itself to the group once they reached the hill, after observing a mysterious pulsating red light nearby: "The [flash]light revealed a towering 'man-like' figure with a round, red 'face' surrounded by a 'pointed, hood-like shape.' The body was dark and seemingly colorless, but some would later say it was green, and Mrs. May reported drape-like folds. The monster was observed only momentarily, as suddenly it emitted a hissing sound and glided toward the group. Lemon [the seventeen-year-old] responded by screaming and dropping his flashlight, whereupon everyone fled."[18] Subsequent investigations at the time of the incident by journalist Lee Stewart Jr. of the *Braxton Democrat* revealed the presence of mysterious skid marks at the site, and UFO writer Gray Barker wrote in an article published the following year that several other people saw the streaking light that had alerted the children and Mrs. May, concluding that the evidence was in fact consistent with the inference that a flying saucer had landed near Flatwoods that night.

Joe Nickell reopened this cold case decades later, when he had a chance to visit the location and talk to some of the people involved who were still alive. There is no doubt in Nickell's mind that this was not a hoax, but his clever and painstaking investigation (he is the Sherlock Holmes of paranormal claims) reconstructed the story from a less extraterrestrial perspective, while accounting for every detail reported in the press or recalled by witnesses. The first thing that became clear is that the UFO was in fact a large meteor, which had been observed across Maryland, Pennsylvania, and West Virginia. Indeed, according to ninety-five-year-old Johnny Lockard, one of the original witnesses, most people had recognized the phenomenon for what it was, and the Maryland Academy of Sciences confirmed the passing of a meteor that night. The celestial body was, however, large enough and low enough on the horizon that some people thought it was a crashing airplane, so much so that the local sheriff had actually investigated the alleged crash site, without finding any evidence of the plane. As for the red pulsating light observed by the children just before seeing the monster, it likely was an airplane alert beacon positioned nearby on the hill (whose light was, in fact, both red and pulsating). The skid marks on the ground found the day after the alien apparition were also easily explained: Max Lockard (Johnny's son) told Nickell that he had explained to the journalist back in 1952 that the marks had been left by his own (Lockard's) Chevrolet pickup truck, when Max had been on the scene looking, ironically, for physical evidence of the apparition.

All of this makes sense, but what about the monster itself? Before reading the answer you may want to go back to the description I transcribed above to see if you can guess correctly (hint: remember that we are in a rural area of West Virginia; think of what sort of things may have been *naturally* going bump in the night). Nickell's reasonable conclusion is that Mrs. May and the kids had actually seen a barn owl! Many of the characteristics of the animal fit the details reported by the eyewitnesses, especially when one considers that they were primed to expect something strange and were positively scared by the circumstances. Nickell notices that there was essentially no description of the alien from the waist down, the most logical explanation for which is that there was *no* waist to be seen. Barn owls do have a strange (and frightening, given the right conditions) head and emit hissing calls and screams, especially when they perceive a threat, such as the one posed by a group of humans suddenly approaching with a flashlight. The owl's behavior can also easily account for the bobbing head and the gliding toward the witnesses that was reported by the group, as well as for the clawlike hands possessed by the alien (they were actual claws; the owl is a raptor bird). Even the local environment is consistent with the habitat preferences of the barn owl, an animal that probably nobody in the group had seen at night in its natural surroundings. To put it as Nickell did at the end of his investigative report: "if it looked like a barn owl, acted like a barn owl, and hissed, then it was most likely a barn owl."

What is interesting abut the Flatwoods case is that it was still possible to find a reasonable solution decades after the facts, which, however, also highlights one of the chief reasons why there are many "unsolved" UFO mysteries: except for Nickell and a handful of others, nobody has the time and resources (not to mention the skills) to patiently investigate UFOs case after case, so that the true believer will always be able to point to unexplained occurrences. But as we have seen before, unexplained means just that—that we do not have an explanation, and there are several mundane reasons for why explanations may be lacking in any particular instance. The other lesson to be learned from the Flatwoods incident rests on what it means in these cases to "explain." A ufologist could reasonably object that Nickell has not *proved* that there were no aliens on that hill back in 1952, that he has only proposed an alternative scenario to account for the known facts. This is true, and it is typical of much scientific investigation, not to mention of everyday forensics—a good model for what Nickell and other skeptics engage in. It is hard, if not impossible, to "prove" a theory, certainly not in the same sense in which we prove, say, a mathematical theorem or a logical proposition. In forensics, science, and paranormal investigations, the best one can do is to have different hypotheses *compete* against each other for the best explanation of the available data. What constitutes "the best" explanation is something that philosophers have been spending quite a bit of time pondering, and it is a question that itself does not admit of an easy answer.[19] However, one of the criteria for a good explanation is that it requires the least number of unwarranted assumptions, something that philosophers call Occam's razor. This is the meaning of Nickell's conclusion along the lines of "if it looks like an owl . . ." The alternative is that, while it looked and behaved like an owl, it was really an extraterrestrial visitor who stopped by Flatwoods, West Virginia, for no apparent reason at all. Use your common sense and see which side you end up betting on.

The last example is meant to teach a lesson to the skeptics, not because it is a "true" UFO (whatever that might mean), but because it underscores the fact that skepticism and critical thinking often require legwork on the part of the practitioner. One cannot simply sit down and shake one's head from a position of alleged intellectual superiority. The facts as we know them once again point to a real event: on 5 March 2004, a reconnaissance aircraft of the Mexican government was flying over the states of Campeche and Chiapas, looking for evidence of drug smuggling. The crew videotaped the appearance of up to eleven unidentified objects that appeared to be visible only in the infrared (which, given the unfavorable meteorological conditions, was the first clue to what the UFOs really turned out to be). Despite initial claims to the contrary, a local radar installation could not, however, verify the sighting (the radar did pick up something that night, but neither the number nor the position of the objects corresponded at all to those of the infrared lights—another clue to what happened that evening). This is a well-documented case with reliable government sources and a video for anyone to examine. Again, something happened over the skies of Campeche, but what?

There was an unfortunate rush of half-baked skeptic "solutions," recounted in disconcerting detail in an article by investigator Robert Sheaffer.[20] These included meteor fragments (according to Jose de la Herrin, a local astronomer), electrical flares in the atmosphere (courtesy of the imagination of Dr. Julio Herrera of Mexico's National Autonomous University), sparks of plasma energy (Rafael Navarro, also of the National Autonomous University), and a group of weather balloons (the explanation offered by the Urania Astronomical Society of Morelos). What these explanations have in common is that they are all dead wrong and that they were all advanced without much thought by someone who simply *knew* that there could not be any unidentified flying object at all. They also clearly display the downside of skepticism: it can easily turn into an arrogant position of a priori rejection of any new phenomenon or idea, a position that is as lacking in critical thinking as the one of the true believer, and that simply does not help either science or the public at large. It turns out that the Campeche UFOs were actually stationary objects on the ground, over the distant horizon: they were flares erupting from a group of oil wells (hence their bright infrared appearance), and the crew of the airplane became confused as to what was moving relative to what else because of the distance, the unfavorable atmospheric conditions, and the unfamiliarity with the phenomenon. Still no evidence of extraterrestrials then, but disturbing evidence that even credentialed skeptics cannot necessarily be trusted to have done their homework.

Ufologists do have arguments that they adduce in favor of their plea for the field to be considered a real science. Yet these arguments are not only unconvincing, but fraught with logically fallacious reasoning. According to James Oberg,[21] the classical arguments advanced in defense of ufology include:

- *Appeals to authority*, as in "Jimmy Carter saw a UFO" or "astronauts saw UFOs" and similar.[22] The problem with this argument is not just that an appeal to authority is intrinsically questionable as a source of evidence, but that often said authorities have no expertise in atmospheric or astronomical phenomena (certainly Carter does not, and even astronauts or airplane pilots may have experience of *observing*, but not *investigating* such phenomena).

- *Affirming the consequent*, another logical fallacy, as in—to use Oberg's apt example—the universe is so large that there must be other civilizations out there, so there are indeed other civilizations, and some of them are visiting us. First of all, it simply doesn't follow from the size of the universe that there *must* be other life forms in it. This may be likely, depending on what assumptions one makes about life and the composition of our galaxy, but it is very difficult to make reasonable estimates in this area. Second, even if there are other civilizations out there, that is no argument at all to think that interstellar travel is happening, or even that it is physically possible. I enjoy *Star Trek* as much as the next guy, but I try to keep a wide dividing line between science fiction and science fact, or even science speculation.
- *Bandwagon appeal*: since so many people believe in UFOs, there must be something to it. But of course widespread belief is no scientific argument at all, since many people believe all sorts of bizarre and unfounded things. Just remember that for a long time the majority of the earth's population sincerely believed that our planet was at the center of the universe. They were all wrong. Moreover, the three cases briefly discussed above make it clear why so many people may sincerely believe that they saw a UFO while in fact they were mistaken.
- *Conspiratorial appeal*: the government knows, but they will not tell us. This is directly analogous to the sort of conspiracy theories we have seen killing people in Africa in our discussion of HIV-AIDS earlier in this chapter. Unfortunately, governments all over the world, including that of the United States, do have a poor record in leveling with their people, despite much talk about democracy and freedom. But, again, while this is reasonable ground for skepticism about what a government says (or does not say), one needs *positive* evidence before scientists can seriously consider a new phenomenon. Besides, the likelihood that conspiracies succeed is inversely proportional to how many people are involved in the conspiracy. If we are talking about something that is allegedly being hidden by most governments in the world, and has been so for decades, the chances of a single significant leak are so high that it is much too high a tribute to the cunnings of governments to think that they would be able to hide such a momentous discovery so well.
- *Salvational appeal*: this brings us back to where we started, cultlike groups such as Heaven's Gate. A recurrent pattern in ufology is the idea that somehow the extraterrestrials are here to help us, to save us from self-annihilation, or to show us the way toward a new path of enlightenment. It is at this level that ufology

crosses the path of religion, something that it is not likely to help its chances of ever becoming a respectable science.

Why is ufology so attractive to so many people? Oberg's answer could apply equally well to the fascination for astrology, as we have seen earlier, or for any other pseudoscience: he thinks of it as "a protest movement against the impersonality and specialization of modern science" and an ego trip of sorts (the insiders know better than most other people in the world, including so-called experts with Ph.D.s). I can sympathize with this feeling of protest, as I recall being attracted by UFOs and paranormal phenomena when I was a teenager. The way I eventually outgrew that youthful phase was through appreciation of science and the real power that it gives us to understand the world as it is. Would I wish for a flying saucer to land on the lawn of the White House to make first contact? You bet. Do I think it likely to happen? Not really, though a good skeptic always keeps the possibility open. Live long and prosper . . .

The Scientific Investigation of Paranormality

I simply could not write a chapter on pseudoscience and not touch, however briefly, on the world of extrasensorial paranormal phenomena. Just as in the case of ufology, however, this section could easily be expanded to an entire book, as much has been written on the topic already.[23] Instead, I will focus on two of the few examples we have of scientific study of parapsychological phenomena to see how the field has fared after more than a century of research into claims of telepathy, clairvoyance, telekinesis, and the like.

"I don't know why this took up a whole lifetime," said William Happer, a physicist at Princeton University. He was talking to a *Nature* magazine reporter about the forthcoming closing—after twenty-eight years—of PEAR, the Princeton Engineering Anomalies Research lab, one of the very few academic laboratories ever dedicated not to engineering, as the name somewhat misleadingly implies, but to research on paranormal phenomena.[24] Then again, the people running PEAR, which was established in 1979 by Robert Jahn (a former dean of the School of Engineering) with initial funding from James McDonnell (of the airplane industry), have always had a way with terminology. Jahn and colleagues probably realized that talk of psi would not go down well in the ivory tower, so they decided that the subject of their studies was "the interaction of human consciousness with sensitive physical devices, systems, and processes common to contemporary engineering practice." In other words, telekinesis.

What the PEAR group did surely qualifies as science, as it is little different in methods from the sort of research that is carried out every day in laboratories around the world: the group's members had an idea, that the human mind can somehow interact with otherwise random processes and bias them in a particular direction; that idea is eminently testable, since it is possible to build devices that allow scientists to conduct properly controlled experiments. Indeed, the idea has been tested, over and over, and the results strongly point toward the conclusion that it is wrong: there is no demonstrable "anomalous transfer of information" between mind and matter.

This isn't exactly what PEAR reported, however. In fact, Jahn told *Nature* that he was shutting down the lab because he was satisfied with what he had originally set out to do. He is convinced that the results accumulated over literally millions of trials unequivocally show mind-matter interaction, because his data indicate that people can "force," say, a machine-generated sequence of random numbers to deviate from expectations on average two or three times every ten thousand. In other words, PEAR claims to have found a statistically significant effect of a size of about 0.00025. This is a rather tiny effect, but if it were true it would still cause a revolution in the way we think of the basic behavior of matter and energy. And yet, the closing of PEAR has been greeted with a sigh of embarrassed relief, not regret, by those few scientists who were paying attention at all. What is going on?

The PEAR results can be explained without invoking the paranormal because of two well-known problems with this sort of experiment: the difficulty of designing machines capable of generating truly random events and the fact that statistical "significance" is not at all a good measure of the importance or genuineness of a phenomenon. As I mentioned above, PEAR's experimental strategy was to test the possibility that the human mind can cause deviations from the expected outcome of a random phenomenon, like a computer generating random sequences of numbers. The problem is, as any computer programmer will readily testify, that it is exceedingly difficult to obtain true randomness from a machine. Even though plenty of computer software is based on so-called random-number generators, in reality these are sequences that are produced by a given algorithm and that are in fact only an approximation of randomness. However, since the algorithms are good enough for most applications, why would they be a problem for PEAR? Because of the second issue, statistical significance.

There are several ways to statistically analyze data, and despite Mark Twain's quip that there are "lies, damned lies, and statistics," statistical analysis is a vital tool in much scientific research. But just as with any tool, one needs to be aware of its proper domain of application and inevitable pitfalls. The most common type of statistical testing—the one adopted by the PEAR group—relies on a quantity called a "p-value," which is an indication of "significance." One way to understand what a p-value says is to think of it as the probability (given certain assumptions) that the observed data are due to chance, as opposed to being the results of a nonrandom phenomenon. The smaller the p-value, the less likely the data are to be compatible with the chance hypothesis, and therefore the more one is authorized to conclude that something was going on (though, of course, the statistical test per se doesn't tell the researcher *what* was going on). This would seem like a perfect match: after all, what PEAR was interested in was precisely a measurement of deviations from chance caused by a human mind.

The problem is that p-values have a well known and somewhat fatal problem in cases like these: if the sample size—the number of trials, in this case—is very, very large (and in the PEAR case, it was huge), one is guaranteed to find artificially low p-values, indicating a statistically significant result, even though there is nothing going on other than small biases in the experimental apparatus. To understand this, imagine you are flipping a coin a large number of times. If the coin is *perfectly* fair, then your chance expectation is to get tails 50 percent of the time and heads the other 50 percent. On average, this is what you'll get. But if the coin is even slightly imbalanced, because of unavoidable asymmetries in the manufacturing process, then you will get a very slight tendency for, say, heads to come up more often than tails. Suppose that you throw the coin millions of times, and that the overall frequency of heads turns out to be 50.00025, with the converse frequency of tails being 49.99975. This difference is tiny (you may have noticed that it is exactly the result obtained by the PEAR group), and yet the associated p-value is vanishingly small, indicating high statistical significance, because the number of trials is so large. But consider for a minute: facing such results, would you conclude that: (a) your mind interfered with the coin tossing, causing heads to turn up slightly more often than tails (this is what you were trying to do to begin with); or

(b) the coin is not exactly balanced, so that the apparent deviation from randomness is actually an indication that one side of the coin (heads) weighs slightly more than the other? I will bet on (b) any time, especially if the alternative seems to violate the known laws of physics!

Of course, there are very real scientific phenomena whose effect size is tiny, for instance in quantum mechanics, so the above reasoning—while casting serious doubts on the grand claim that PEAR discovered a hitherto unknown scientific phenomenon—is not conclusive. But there are two more pieces of damning evidence to be considered. First, skeptical researcher Stanley Jeffers has taken a closer look at some of the published claims by the Princeton group, focusing in particular on the so-called baseline data, that is, on the controls ran to make sure that the random generator was, in fact, random.[25] Jeffers quotes Jahn and coauthor B. Dunne who, in their 1987 book, appropriately entitled *Margins of Reality*, write that "of the seventy-six baseline series performed, seven or eight of the means would be expected to exceed the 0.05 terminal probability criterion [the threshold p-value of statistical significance], in one direction or the other, simply by chance. In fact not one of them does." This somewhat abstruse language signifies, as Jeffers drily observes, that the control data was too good to be true, which may mean one of two things: fraud or a systematic bias in the control. Unlike in so many other cases of paranormal investigations, there is no reason to believe that the PEAR group knowingly altered their data. But an untrustworthy baseline puts them in the same exact position as our hypothetical coin-flipping experiment above, because if the control is not perfectly random, then tiny observed deviations from the control are not a reliable indication of anything mysterious at all.

Second, to their credit, acknowledged by skeptics alike, the PEAR group has engaged other researchers in different laboratories to attempt to reach the golden standard of scientific experimentation, the replication of their results. Two such independent groups, both based in Germany, have in fact failed to replicate the PEAR results, yet another indication that the simplest hypothesis is likely to be true: there was nothing to replicate.

Lack of replication also plagued the other long-running academic lab devoted to studies of the paranormal, the one established by Joseph Banks Rhine at Duke University (it also has since been disassociated from the university), which constitutes our second example in this section. Rhine was a botanist by training but became interested in parapsychology after hearing a lecture by Sir Arthur Conan Doyle, the author of the *Sherlock*

Holmes books. Doyle was convinced of the possibility of talking with the dead and of the existence of fairies. Although this may seem rather surprising from the same man who gave us the quintessential fictional icon of logical rigor, Doyle thought his beliefs were based on empirical evidence: after all, he had participated in séances, and he had examined photos of fairies![26] Rhine was similarly attracted by a variety of fringe notions, as he devoted some time to conduct research, for instance, on Lamarckian inheritance (so-called acquired characteristics induced by the environment on living organisms), even though the idea had not been taken seriously since the passing of the nineteenth century (and in fact had been rejected by most scientists even before Darwin). For that matter, Rhine—like Doyle—was prone to endorse the downright bizarre, as when he declared Lady Wonder, the "talking" horse, "the greatest thing since radio!"

Rhine is perhaps most famous for his experiments on telepathy involving a set of five cards with geometrical symbols, the so-called Zener cards (after Karl Zener, an associate of Rhine). The idea was simple: mix the cards at random, have an operator look at one card at a time and try to convey the information mentally to a subject, then count how many hits and misses you get. Since there are five cards for each of the five symbols, for a total of twenty-five cards, any statistically significant deviation from the expected chance score of 1/5 would represent evidence of paranormal activity. Well, not exactly.

To begin with, you may have noticed the reappearance here of the same problematic concepts that damned the PEAR attempts: "randomness" and "statistical significance." Rhine's group was plagued by the same sorts of issues, though for slightly different reasons. First, the shuffling of the cards was often done manually, not by a machine, and indeed in several cases it was the allegedly telepathic subject himself who did the shuffling. Now, would it really come as a surprise if I told you that Rhine's highest-scoring subjects suddenly saw their success rate recede toward the chance expectation as soon as they were not the ones handling the cards, or when they were observed by a magician with expertise on card tricks?

Statistical significance was a problem for Rhine for the same reason we have already encountered: when one runs a very large number of experiments, one is bound to find significant (but very small) results even though there is nothing paranormal going on, as long as there is even a very small bias in the experimental setup. But in Rhine's case the situation was made worse by the fact that the statistical theory he used was meant to deal with a "universe" of possibilities much larger than the twenty-five cards actually used. With such small numbers at play, the statistics do not perform well, just like any tool that is being used for something it wasn't designed for (try driving a nail into a wall using a shoe instead of a hammer: it can be done, some of the time, but the outcome is not likely to be polished, and your shoe will be ruined in the process). More importantly, many of the experiments at the Duke lab were carried out using a "no replacement" protocol, meaning that each card in turn was discarded from the deck, not put back. This protocol means that as the test progresses, the probabilities of a given card to turn up *change in a semi-predictable fashion*, like any half-decent card player knows well. If you know, for instance, that four of the five circles have already shown up, the likelihood that the next card will be a circle is now pretty low (unless there are very few cards left in the deck). Such probabilities are called *conditional* in statistics, but the 5/25 chance expectation on which the tests were predicated is valid only with unconditional probabilities, as in the case of a continuously reshuffled deck. There are ways to control for this problem, of course, but it is not at all clear whether proper protocols were always followed in Rhine's lab.

Rhine, to his credit, also tried to have his results duplicated by other laboratories, as did the PEAR people—but to no avail: researchers at Brown University, Colgate, Johns Hopkins, Princeton, and Southern Methodist tried, but they could not get any significant evidence of paranormal phenomena. Most damning of all, Rhine admitted publicly that he had uncovered at least twelve instances of dishonesty among his researchers in a single decade, from 1940 to 1950. However, he flaunted standard academic protocol by refusing to divulge the names of the fraudsters, which means that there is an unknown number of published papers in the literature that claim paranormal effects while in fact they were the result of conscious deception.

Perhaps the most appropriate question for us in this chapter, which can be applied to all the pseudoscience we have briefly examined, is the one that *Nature* reporter Lucy Odling-Smee asked in the piece on the closing of the PEAR lab: "how permissive should science be of research that doesn't fit a standard theoretical framework, if the methods used are scientific?" In other words, how many times do we have to show that alleged instances of telepathy, clairvoyance, telekinesis, ufos, ghosts, psychic abilities, astrological forces, and the rest of the shebang can better be explained by perfectly normal means? There is, of course, no simple answer to this question, as the appropriate amount of resources and time to devote to the investigation of fringe science depends on how important the claims would be if they turned out to be true, how many times they have been disproved before, and, frankly, how limited the resources of scientists and universities are in practice. We hear repeated calls from ufologists, parapsychologists, psychics, and astrologers to keep an open mind, and there are certainly examples of skeptics who close the door to further investigation a bit too quickly, as in the case of the Campeche UFOs recounted above. But, as astronomer Carl Sagan once aptly put it, you do not want to keep your mind so open that your brain is likely to fall out.

CHAPTER FOUR

Blame the Media?

Just because your voice reaches halfway around the world doesn't mean you are wiser than when it reached only to the end of the bar.—Edward R. Murrow, journalist

I find television very educating. Every time somebody turns on the set, I go into the other room and read a book.—Groucho Marx

"The medium is the message," social commentator Marshall McLuhan famously said. Well, not really. What happens is that a given message may be repackaged in highly distinct ways depending on which medium it is broadcast through, and the medium affects how well the message can be communicated. If the message is science, it turns out that it is hard to get it across properly, regardless of whether the medium is a newspaper, television, film, or the Internet; conversely, the same media seem to lend themselves to quickly and effectively propagating all sorts of pseudoscientific nonsense, though some more so than others.

In this chapter we will discuss the relationship between modern media and pseudoscience, and we will learn that our baloney detector needs to be turned on even when we are watching credentialed skeptics tackling certain politically loaded topics. The issue of scientific miscommunication through the mass media is incredibly complex, as the type of media and the sheer number of outlets multiplies vertiginously every year.[1] Indeed, any serious attempt to explore the topic thoroughly would easily balloon this chapter into a full-fledged book. Instead, as I have done in similar situations throughout the book, I will focus on a small number of case studies with the objective of drawing what I believe are some generally useful lessons. It is a treacherous landscape that we are about to explore, characterized by allegedly serious journalists who behave like comedians, as well as by real comedians who could give lessons in serious journalism. It's just that crazy out there.

Show Time! Witches on TV and Celebrities' Nonsense

I always found the concept of "reality TV" rather self-contradictory, and I must admit that I have never been taken by any of the many popular shows of this type that have appeared at a breathless pace during the past several years. But in July 2008 the Australian Channel 7 surprised me by taking the next apparently logical step in that strange realm: they announced a quest for Australia's top psychic![2] Since I don't believe there is such a thing as a psychic, this promises to be the quintessentially oxymoronic reality show: one about a reality that does not, in fact, exist. The idea is to employ two judges, Stacey Demarco, billed as a "practicing witch," and Richard Saunders, vice-president of Australian Skeptics. Guess who is going to take on what role in this paranormal incarnation of the good cop-bad cop routine. Of course, in line with the populist appeal typical of reality-show television, the ultimate judge will be the public, who will choose which of the remaining three contestants is "really" the top psychic.

This, naturally, *assumes* that psychic phenomena are real, a rather questionable starting point. Yet Saunders, for his part, sees this as an educational opportunity for the Australian public: "My role is being a sceptical [*sic*] judge, someone who can size up the contestants and give a rational explanation for what they do. I'm a pretty tough judge because I'm looking for definite psychic or paranormal happenings," he said in an interview to promote the show, "Normally on these shows they have a token sceptic, and I've been one. One of the things about this show is that I get to put my point of view." In other words, Saunders sees this as a rare opportunity to have the skeptical perspective on pseudoscience play more than the usual perfunctory role, and I admire his optimism and willingness to try. Demarco, for her part, commented that "I'm a rational type of expert, I'm not the purple tie-dye type of witch. I just want people to come into this with a really open mind," which sounds reasonable enough: a rational non-purple witch ought to be a guarantee of seriousness for the whole enterprise.

The reason I bring up Australia's quest for the top psychic is because it provides us with a typical example of a conundrum faced by every scientist or skeptical investigator who is genuinely interested in engaging the public in the name of science education: how far should one go? Saunders presumably realizes that he will be on the losing side as far as the majority of the public is concerned: if the show is going to be successful it will likely be because people want to see the psychic in action, not the naysayer who debunks him. Nevertheless, Saunders is betting that playing a role on equal footing with a witch will help the cause of critical thinking. This is not different from the choice I had to face when I pondered the pros and cons of debating creationists, something that I eventually ended up doing on a variety of occasions. Colleagues like Richard Dawkins are adamant that one should not "lower" oneself to that level, because this legitimizes and gives a platform to the pseudoscientist. There is something to be said for that attitude. Conversely, however, a scientist refusing to engage in public debate may also easily come across as the stereotypical intellectual who feels too superior to the masses to bother entering the arena (not that anyone would aim such an accusation at Dawkins, of course). Besides, creationists and psychics already *have* a platform, and it is the scientific and skeptical communities that constantly have to fight for a tiny spot in the limelight.

Eventually I ended up making the same decision as Saunders, and I have only occasionally regretted it. To begin with, a public appearance of that type is a cultural opportunity hardly to be missed: I learned a lot more about creationism as a social phenomenon, and about creationists as people, than I would ever have been able to glean from reading articles and books about them. I also quickly realized that being prepared (as in being knowledgeable in my technical field) is a necessary but by no means sufficient condition. Scientists who engage in public appearances, and particularly debates or similar events, typically make two mistakes, both fatal (metaphorically speaking). The first error is to assume that because the opponent is a pseudoscientist it will be easy to dispatch whatever argument or objection he may come up with. This grossly underestimates the complexity of the pseudoscientific culture, where an honest scientist can suddenly find herself facing the most bizarre questions on the most arcane topics, well outside of her direct technical expertise. For instance, during the question-and-answer section following a public debate with creationist deacon Duane Gish, I was asked by a member of the public how I reconciled evolution with what we know about basalt. Basalt? What does that have to do with anything? Not much, except in creationist logic.[3] It turns out that according to a well-known "theory" proposed by the Institute for Creation Research, of which Gish is a member, the basalts found in the Columbia River are geologically very recent and were formed during the worldwide flood that involved Noah and his famous ark. Had I not done my homework and read what serious geologists have to say about these formations (they originated three and a half million years ago, as a result of plate tectonics), I would have been staring into the void like an idiot, confirming the audience's suspicion that the so-called expert didn't actually know what he was talking about.

The second mistake that many scientists make under these circumstances is to treat them as academic debates, where positions are evaluated on the basis of evidence and rigorous logic, let the best contestant win the day. Rather, the appropriate comparison is with political debates such as the presidential ones in the United States, where one "wins" not (just) because one may have the better arguments, but because one is better at wielding rhetorical tools, comes across as a nice fellow, or, simply, is better looking than the opponent (do not laugh—this is what famously tilted the scale to John F. Kennedy's side during a debate against Richard Nixon, with people listening to the radio declaring Nixon the winner, while people who watched it on television decided the other way around apparently because they were enthralled with JFK's charm and disliked Nixon's copious sweating). Several times after a debate I have been approached by creationists who thanked me for being nice and reasonable and who commented that I seemed like an amiable enough chap, not one of those baby-eating, fire-spitting devils that they expected from the descriptions of scientists made by their minister! So I sincerely hope that Richard Saunders will fare well in his attempt to make Australians question the very idea of psychics and witches, though it will be an uphill battle from the beginning.

There is a related phenomenon that also connects entertainment and bad science, but in a different, and perhaps more insidious, manner. One of the reasons reality-television shows are popular is because we live in a culture of celebrity, where being famous, for whatever reason (or for no reason whatsoever), seems to be the goal of millions of people. Andy Warhol is well known for having said that in the future we will all have fifteen minutes of fame, and he himself was a good example of how the mass media can be manipulated to that end. Reality TV and even more so the Internet are putting such a nightmare within reach of our generation.

As far as science in particular is concerned, a special problem arises when celebrities make public statements about scientific matters even though they have no science background whatsoever. For instance, actor Richard Gere endorsed crystal therapy to help in adjusting the body's natural frequencies (whatever they are), while another actor, Tom Cruise, affiliated with the Church of Scientology, lashed out publicly against psychiatry, claiming that "there is no such thing as a chemical imbalance [of the brain]." Robert Kennedy (the son of assassinated presidential candidate Bobby) told *Rolling Stone* (not a scientific journal!) that a vaccine causes autism and moreover that the government has attempted to cover up this invented truth. Former Beatle Paul McCartney's former wife (I know, it gets complicated, but bear with me) Heather Mills linked child obesity to drinking too much milk (studies have been carried out, and there is no connection), and Prince Charles of England advocates a mixture of coffee, fruit juices, and vitamin injections as a cure for cancer.

The list could go on for pages upon pages, but these examples will suffice to make the point that it is *positively dangerous* for people who have no science background to talk in public about scientific notions that they obviously do not understand. Particularly so when these people enjoy an unusually high level of visibility and public trust, however ill gotten. This is why a British organization, *Sense About Science,* has taken the most welcome step of publishing a leaflet to help celebrities speak properly about science, if they really must.[4] A celebrity who wishes to do some fact checking can even call a special hotline made available for just such a purpose; the number is 020–7-478–4380 (don't forget England's code of 44 if you are calling internationally). Had she called, perhaps "lifestyle guru" Carole Caplin would have spared herself the embarrassment of going public with the ludicrous idea that women need to keep their lymph system "clear and unclogged" to avoid breast cancer, a notion that *Sense About Science* real expert Michael Baum of the University College in London commented "has no meaning whatsoever." But enough shooting fish in a barrel; the point, I think, is clear enough.

What seems to me astounding in these episodes is the dangerous combination of a gullible public and a group of high-visibility people who apparently display not the least amount of shame about blabbering nonsense—precisely the same combination we have encountered in the last chapter when we talked about AIDS denialism, a phenomenon that is literally killing millions. We live in a culture where real experts are derided or ignored, and yet where we are willing to spend money, and sometimes stake our lives, on the advice of people who manifestly do not know what they are talking about. If there ever was a reason to push for more education not just about science, but about basic critical-thinking skills more generally, this seems to be it. Then again, perhaps not all is lost: the news media—at least in open societies—are supposed to play a crucial role in educating the public, mediating between the decision makers and experts on the one side and the everyday women and men that make up the fabric of our society on the other. There certainly are excellent examples of this kind of media function, for instance the articles on AIDS denialism published in the *New Yorker*, which I cited in the last chapter. But it only takes a quick glance at most of the output of generally respected media outlets to begin to have some doubts.

Newspapers, Television, and Mass Bunk in the Mass Media

On 19 June 1998 the *Washington Post*, one of the most respectable newspapers in the United States, wrote: "the first independent scientific review of the controversial topic [of UFOs] in almost 30 years [found] cases that included intriguing and inexplicable details, such as burns to witnesses, radar detections of mysterious objects, strange lights appearing repeatedly in the skies over certain locales, aberrations in the workings of automobiles, and irradiation and other damage found in vegetation." As Robert Sheaffer wrote, the *Post* wasn't the only media outlet to uncritically report about the new "scientific review," as both the Associated Press and ABC News were taken in.[5] It should be clear by now what I think about ufology in general, but what makes this case particularly disturbing is that it would have taken a minimum of investigation on the part of the reporters to figure out what was going on.

As it turns out, the so-called independent scientific panel was actually a charade put together by the Society for Scientific Exploration, which is not a mainstream scientific organization at all, but rather a group of people—including some scientists—who are sympathetic toward paranormal beliefs and UFO claims. One of the panelists, for instance, was longtime ufologist Jacques Vallé, who has often advanced the very strange argument that there must be something really important about UFOs *because* the phenomenon makes no sense at all! Had the journalists at the *Post* or ABC bothered to do a few minutes of Internet searching and perhaps ask a couple of questions on the phone, they would have also found something interesting about the chairs of the panel: Harold Puthoff is the director of Austin's Institute for Advanced Studies, but is also a well-known parapsychologist and conducts research on so-called zero point energy, the idea that one can extract energy from empty space—a proposition, I should add, that violates basic principles of thermodynamics and that is considered pseudoscience by credentialed physicists. The other chairperson for the "scientific" panel was David Pritchard, a physicist at the Massachusetts Institute of Technology who claims that alien abductions are real phenomena that ought to be considered seriously by mainstream science, despite the utter lack of evidence and the availability of perfectly normal explanations for the cases that have been seriously investigated. Moreover, the instances trumpeted by the panel as "new evidence" were in fact well known in the ufological literature and had already been properly debunked.

What happened was that otherwise serious media outlets were completely taken in by a well-written press release and by the presence at the conference of a couple of legitimate scientists. The latter is certainly not a sufficient excuse for not digging deeper. As we have seen time and again, it is always possible to find academics with sound credentials who will espouse all sorts of bizarre ideas, often with genuine conviction. There are scientists who are creationists, others who deny that oil is a fossil fuel, and those who do not think that global warming is happening; there are scientists who believe in telepathy and clairvoyance, and those who deny that the HIV virus causes AIDS. This does most certainly not mean that such matters are really open to fundamental doubts at this point (although some of them may have been in the past), it just means that having a Ph.D. and working in a university is neither necessary nor certainly sufficient to provide one with unquestionable authority, no matter what the subject. A recurring theme of this book is that one cannot simply trust authority no matter how, well, authoritative it may appear to be. There is, unfortunately, no shortcut to using one's brain and critical sense and doing some background research before taking a position. If this is true for all of us as private citizens, it is a fortiori true for the news media, which after all have a special ethical responsibility to inform the public in the best way they can.

One of the doctrines adopted by many news media during the past several years, and which has been much criticized by scientists, among others, is the idea that there are often two sides to a given story and that they ought to be equally represented as a matter of fairness. Even a moment of reflection will readily show the flimsiness of such a "fair and balanced" treatment: first of all, in many instances of scientific (not to mention political) debate there actually is a more nuanced landscape than a simple yes or no dichotomy. For instance, in the case of global warming even the best scientific models available produce a variety of possible outcomes, about the likelihood of which there is intense and legitimate debate. Second, however, it should also be obvious that not all alternative positions are equally deserving of public attention and that therefore they should not be presented as equivalent opinions about which "you decide." There is no scientific controversy about evolution versus creationism, and the fact that half of the American public rejects the scientific findings in this area is an interesting, and worrisome, social phenomenon, but certainly not a measure of scientific uncertainty! If the news media are to play a truly informative role with the public they should present more than just a collection of allegedly equally valid ideas; they should also do the hard work of investigating them, to help the public filter the few golden nuggets from the ocean of nonsense that will otherwise bury any intelligent social discourse.

And yet, as a scientist and skeptic, I actually wish that the news media would in fact stick at least to their stated policy of equal opportunity, because that would be a great improvement on the actual state of things, where the rational viewpoint is given only a passing nod and is otherwise buried by sensationalism and sloppy journalism. Take, again, the treatment by mainstream media of the UFO phenomenon, as analyzed by author Philip Klass on the basis of his repeated first-person experiences.[6] Larry King of CNN fame, for instance, devoted a TNT network two-hour special to the infamous "Area 51," the secret Nevada location where the U.S. government has supposedly stored a crashed flying saucer, as well as the bodies of several aliens (as depicted in the documentary, I mean movie, *Indiana Jones and the Kingdom of the Crystal Skull*). Klass points out that King attempted to provide viewers with the illusion of a balanced treatment by inviting Klass and astronomer Carl Sagan as the token skeptics on the program. Out of the two hours broadcast, however, Klass was on screen for a total of ninety seconds, and Sagan a whopping seventy-five seconds. As if that were not bad enough, Klass had given the producers of the show some never-before-released, previously classified air force documents providing new evidence (if any was in fact necessary) that no flying saucer had crashed in New Mexico, and could not therefore be stored at Area 51. Not only did TNT producers completely ignore the documents (bad journalism), but King rather disingenuously concluded the show by saying "Crashed saucers. Who knows? But clearly the government is withholding something . . ." As Klass wryly points out, it was actually Larry King who was knowingly withholding pertinent information from his viewers, thus engaging in egregiously unprofessional journalism.

To make things more difficult, it is not possible to completely trust self-professed skeptics either (and you should not take my word for it!). A generally refreshing outlet for critical thinking is the Showtime television program *Bullshit!* hosted by professional magicians Penn Jillette and Raymond Teller. The show is funny, if often crass because of Penn's tendency to curse, and very intelligently put together. Each episode examines some pseudoscientific claim and proceeds to debunk it with a combination of investigative journalism and empirical demonstrations. Penn and Teller do not pretend to be doing rigorous science—after all, it is a television program meant to entertain—but their antics also manage to educate, and I use them regularly in a class I teach on the nature of science, to provide my students with what turns out to be a very effective combination of laughs and food for serious thought.

Even Penn and Teller, however, sometimes get it spectacularly wrong, and it is instructive to briefly examine one example because it reinforces several of the points I am trying to make throughout this book. Episode 13 of the first season of *Bullshit!* which aired in 2002, tackles the problem of global warming, to which I will return in detail later in the book. The choice of topic may appear strange at first glance, since—despite the scientific discussions and the sociopolitical controversy—atmospheric physics certainly is no pseudoscience. It takes only a few minutes of background research to begin to guess why Penn and Teller are so skeptical of global warming: Penn Jillette is a well-known libertarian and a fellow of the Cato Institute, a think tank that has repeatedly taken positions against the emerging scientific consensus on global warming. The Cato Institute, it should be added, is funded in part by the Exxon-Mobil Corporation, not exactly a neutral player in discussions about energy production and use. Of course, the suspicion of bias is not enough to condemn Penn and Teller's treatment of global warming, but one's baloney detector's alert level should go up a couple of additional notches once a few more things become apparent from the broadcast. First, Penn and Teller pit oil-industry lobbyists against hippie college protesters to make their case that the global warming movement is a sort of New Age irrational belief. The only credentialed scientist to speak on the program is economist Bjorn Lomborg, a notorious skeptic of global warming (and not an atmospheric physicist), to whom I will devote much space later on.

Things get worse as the show transitions from a clearly imbalanced presentation to outright *mis*representation of the debate.[7] One of the guests is Jerry Taylor (of the above mentioned Cato Institute), who says "In the mid '70s we were told pollution is going to cause a new ice age . . . The very same scientists who argued an ice age was coming because of industrial pollution then shifted gears and argued industrial pollution will bring on a greenhouse warming world with virtually no breath in between." This is simply false, as the idea of a temporary cooling of the earth's temperature was advanced in the popular press (not in academic, peer-reviewed journals), prompted by speculations about the massive injection of aerosols in the atmosphere. To compare a few magazine articles with the overwhelming scientific literature on global warming is a joke, and not a particularly funny one, given what is at stake. Then again, to make things even less black and white, Penn and Teller do manage to score some good cautionary points about what they see as "environmental hysteria." In a truly hilarious, and in fact quite disturbing, scene, they successfully convince several college students to sign a petition against the use of dihydrogen monoxide, an allegedly terrible environmental pollutant, otherwise known as . . . water.

It is perhaps a sad state of affairs when entertainers like Penn and Teller do a better job than CNN or ABC News at promoting critical thinking among the public, despite the just mentioned faux pas on global warming. It should be positively frightening that one of the most informative and properly balanced recent treatments of the creation-evolution war was done by comedian Jon Stewart.[8] Stewart's *Daily Show* is best known for its biting political satire and its progressive-leaning perspective on world affairs. However, in 2005 the comedian devoted a whole week of his show to what he termed the "Evolution, Schmevolution" issue, because of the then ongoing trial on the teaching of intelligent design in Dover, Pennsylvania. I will cover that momentous trial and its implications in detail later on, but it is worth noting that Stewart and his gang of fake journalists managed to achieve several objectives that one would expect to be the purview of the "serious" media, which instead perform consistently below par on the matter.

To begin with, Stewart got the science of evolution and the claims of intelligent-design proponents exactly right. Moreover, his writers used humor to immediately point out that the two perspectives are not at all equivalent. Here is what regular guest Lewis Black had to say about teaching creationism: "Scientific theory is based on observations made in the real world. . . . [Instead, for] little creationists there is *D is for Dinosaurs*, where kids are taught that before the Flood all dinosaurs are vegetarian. Makes sense, especially when you look at this early dino-skeleton [points to a fossil of a meat-eating dinosaur]: those 80 dagger-shaped teeth and huge claws were perfect for chasing down and killing any plants that tried to run away!" Black manages to convey a sense of how scientists actually arrive at reasonable inferences about the fossil record (if it had long teeth and claws, it was probably a meat eater), while at the same time underlining the fact that the creationist position is not only antiscience, it simply lacks common sense.

As I mentioned above, I have participated in debates with creationists—an interesting cultural experience in its own right, and one in which it is exceedingly difficult for the rationally inclined to make headway. Not so for Stewart. In the last episode of the four-part series on evolution, he convened a panel of three experts to chat with: historian Edward Larson, author of the Pulitzer-winning *Summer for the Gods*, a history of the infamous Scopes monkey trial of 1925; William Dembski, a leading proponent of intelligent design; and a loony character named Ellie Crystal, who was there solely to provide (additional) comic relief. At one point during the discussion Stewart turned to Dembski: "Let me ask you this: intelligent design, the scrotum, the most painful part of my body. This intelligent designer chose to put it in a bag that anyone can walk across and hit with a baseball bat . . ." Dembski realized immediately that, despite the joke, the point was serious and has in fact been made (somewhat less jocularly) by biologists since Darwin: how do intelligent design proponents deal with so many obviously unintelligently put together characteristics of the biological world? Dembski, to my astonishment, conceded the point, something that is much harder for a professional biologist to wrangle from a creationist, admitting that his theory is not committed to explain "everything" and that natural selection must have played a role in evolution. Dembski did not elaborate on the difficulty implied by his response: how exactly does one tell the difference, then, between what was designed and what was not?

During the panel discussion, Stewart highlighted the same problem with Dembski that I pointed out about Penn Jillette above: inherent bias. Stewart candidly asked Dembski during the program: "What came first [for you], the religion [*sic*] conversion or the evidence [against evolution] convincing you?" to which Dembski, to his credit, admitted "the religious conversion came first." Now, as I already said in the case of Jillette's libertarianism and association with the Cato Institute, one cannot *simply* dismiss one's position based on real or perceived biases. For one thing, this is a common logical error, known as the genetic fallacy, and moreover everyone has biases, or even simply opinions, if one digs deep enough. However, note that the fallacy would be to conclude that Demsbki (or Jillette) is wrong *solely* on the basis of his stated ideology; one needs more than that—chiefly, of course, a thorough analysis of the claims made and the evidence adduced. But we are all human beings, and evidence of bias, as we will see when talking more in depth about "think tanks," is a useful warning sign that should raise our defensive shields and prompt us to look particularly carefully at the statements being made. That we need Jon Stewart to remind us of this elementary precaution is one more reflection of the sorry state of public discourse in modern society.

Science and the Meaning of Life: What the Bleep Do We Know?

One does not expect to learn good science from fictional movies, and despite my love for *Star Trek* I do not think that there is a particularly close connection between starship communicators of the twenty-fourth century and cellular phones of the twenty-first century (I mean, in three more centuries we will still have flip phones? C'mon). But when a production is billed as a documentary, things take on a different light, and the viewer has a right to demand a product that has passed the muster of basic fact-checking, not to mention of elementary logic. Of course, much of the time the viewer will be disappointed. One of the most spectacular examples of a horribly tangled mess of science and nonsense is *What the Bleep Do We Know?* released in 2004 and which, at $10 million and counting, is one of the highest grossing "documentaries" ever. We will use it as an example of how not to do a science documentary and as a vehicle to gain further insight into how difficult it is to communicate both solid science and critical-thinking skills.

What the Bleep is part fictional story of a woman trying to make sense of her life and part documentary based on cartoons illustrating scientific knowledge and interviews with a series of experts commenting on the relation between science and the meaning of life. It is only at the end of the movie that we find out that one of the "experts" with the most onscreen presence is none other than Ramtha, a thirty-five-thousand-year-old spirit warrior from the lost continent of Lemuria. Ramtha does not talk to us directly (but why not?), but through the "channeling" of spiritual teacher J. Z. Knight, who reportedly makes a hefty $1,500 per person from people who attend her retreats.[9] Is your baloney detector at least on yellow alert already? Good, then let us proceed.

The central idea of the movie is that the foundation of reality is to be found in thought and consciousness, not matter and energy, and that therefore the best we can do to solve our problems is to simply rethink reality itself. This is not a novel position per se, having been espoused in one flavor or another by plenty of both Western and Eastern philosophers. But the ironic twist presented by *What the Bleep* is that its New Age spiritualism is actually supposed to be grounded in the hardest science of them all: quantum mechanics (with a sprinkle of neurobiology to make the connection to consciousness less obscure). This is actually a recurring theme in much pseudoscience literature, from ufology to parapsychology to creationism: even though the basic idea is to reject (or "go beyond") modern science, which is seen as a highly questionable incarnation of all that is wrong with Western greed and reductionism, pseudoscientists desperately seek to buttress their arguments with science, seen somewhat bipolarly both as an irreducible foe and as a powerful legitimizing force. Fascinating, as *Star Trek*'s Spock would surely say.

What the Bleep is chockfull of non sequiturs and scientific mistakes, as has been pointed out in plenty of commentaries all over science outlets. But this is also one of those rare cases where we can find an exceedingly lucid in-depth analysis by a mainstream journalist, Bernie Hobbs of ABC (that is the *Australian* Broadcasting Corporation, not the better-known American synonymous organization). Let us take a quick look at some of Hobbs's points to see just how well, for once, a journalist can do his job if he just uses a bit of common sense and bothers to check his sources. Several times in the documentary the authors begin with the description of an actual scientific phenomenon and then extrapolate from there to conclusions that do not follow from the premise and that in fact are at odds with our best understanding of how the world actually works. For instance, quantum mechanics tells us that subatomic particles do not exist in the standard everyday sense of "existing." Rather, as we have seen while talking about string theory, their physical properties are in a constant state of flux, described by a range of probabilities known as the wave function, and they "collapse" into specific values[10] only when the particles interact with each other or with a measuring instrument. From this, *What the Bleep*'s authors conclude that reality *as a whole* is in flux and that consciousness is needed to make it stable. But as Hobbs and others have pointed out, this is simply not true: a rock will exist regardless of anything else, conscious or not, interacting with it, because quantum behavior describes what matter/energy does *at the quantum level*, not at the macroscopic level that concerns most of human existence.

Similarly, the documentary makes much out of one of the favorite, and most abused, ideas in quantum mechanics, Heisenberg's uncertainty principle. It essentially says that it is impossible—in principle, not just as a matter of insufficient instrumentation—to accurately measure both members of certain pairs of properties of subatomic particles, such as the momentum and position of an electron. Physicists often refer to this, rather improperly, as the "observer" effect, but they do not mean that an actual human observer is necessary for the effect to take place; they are talking about a particular kind of physical interaction, which has nothing to do with consciousness per se. So to conclude that Heisenberg's principle implies the necessity of a conscious agent in order to literally "make" reality is a preposterous misunderstanding of the physical theory.

Another splendid instance of nonsense highlighted by Hobbs is a statement by one of the featured speakers in *What the Bleep*, Dr. Joseph Dispenza, a chiropractor who apparently has no qualms talking as an expert on a field quite apart from his own specialty: neurobiology. Dispenza, again in the same vein of beginning with sense and leaping into nonsense, claims that brain scans show that the same areas of the brain that are active when one is having a particular experience light up also when the subject is *remembering* that same experience. This is in fact true. From this, however, Dispenza concludes that the brain literally cannot tell the difference between reality and imagination! Besides the simple observation that those same brain scans clearly show that the intensity of brain activity is much lower when one is remembering than when the actual experience takes place (in other words: the brain apparently *can* tell the difference), how can anyone seriously believe that? If Dispenza were right, we would continuously confuse, for instance, dreams and reality. While this is indeed the case for rare pathologies, and many of us experience the occasional nightmare vividly enough to wake us up in a cold sweat, we simply would not be able to function as individuals and societies if our brains were literally incapable of separating reality from imagination. Yet another extraordinary claim based on extraordinarily little evidence.

One of my absolute favorite bits of the movie, however, arrives when one Candace Pert (who is billed as a former scientist and current "guru") says rather matter of factly that Native Americans literally *could not see* Columbus's ships because they could not make sense of such artifacts within their cultural reality. This is a quintessentially extreme postmodernist position (more on this later in the book) in which we do not just construct our reality out of social intercourse, but we literally cannot perceive any reality that is not part of such intercourse, even if we are about to be wiped out by some of the consequences of this allegedly undetectable reality. Pert's claim is so bizarre that one does not even know where to begin showing in how many ways it is wrong. First of all, it would be nice to be told, as Hobbs points out, how Pert knows that pre-Columbian Americans could not see the caravelles. Did she channel the spirit of a fifteenth-century Native American who somehow could explain to her this bizarre reality-negating experience that he and his fellow tribesmen had? Moreover, simply imagine the consequences for us all of Pert's extraordinary assertion: this would mean, essentially, that no culture could ever physically see anything that was not already a part of that culture's discourse. The absurdity of the claim defies belief.

Another protagonist of the movie is Dr. Masaru Emoto, who makes his living publishing pictures of thought-altered water. You read correctly: Emoto claims that one can talk to water and that water crystals take different shapes depending on whether one says nice or mean things to the water. This is important in the scheme of the documentary because, as Emoto puts it, "If thoughts can do that to water, imagine what our thoughts can do to us." Except, of course, that thoughts can't do anything at all to water, unless your thought is to get up and drink some (and you then act on it). Hobbs reports that the magician and skeptic James Randi has offered a hefty prize of one million dollars to Emoto if he can reproduce the thought-on-water effect under controlled conditions. Emoto has not agreed to the experiment, presumably because the royalties from his books represent a steadier and less challenging source of income.

My final example from *What the Bleep* is a famous experiment that was attempted in Washington DC in 1992. In the movie, John Hagelin describes his study that ostensibly showed that four thousand volunteer meditators were able to significantly lower the crime rate in the U.S. capital just by meditating and producing positive vibrations. Were this true, of course, I would be the first to call up my local police department and lobby it for a significant investment in meditation techniques for its officers. It would make for rather boring episodes of *CSI*, but that would be a fair price to pay for less violence in the streets. The problem, of course, is that Hagelin's claim is plainly false: the alleged 18 percent drop in crime rate was relative to Hagelin's own prediction of what *would have happened* that summer in Washington if he and his meditator squad had not intervened. It is not at all clear how Hagelin got his figures to begin with, and anyone else who examined the actual, instead of the imagined, data saw a *rise* in the crime rate in the nation's capital during that summer. And so on and so forth. *What the Bleep* goes on for an excruciating one hour and forty-eight minutes, and that does not include the subsequent six-part DVD set *Down the Rabbit Hole*, which I have to confess I did not have the stomach to watch.

What the Bleep Do We Know? is a particularly worthy example of nonsense being propagated by mass media, in this case in the form of an alleged documentary, both because it contains many of the typical kinds of "reasoning" that fuels this sort of enterprise and, more worrisome, because it has become so successful so quickly, with people claiming that it changed their lives and the way they look at the universe. How unfortunate.[11]

The Altenberg 16: Conspiracy Theories by and for the Media

The last example of media's bizarre treatment of science is rather a tempest in a teapot, as its reach has been confined mostly to a flurry of Internet articles and blog posts over a period of a few months. However, it is a good case study because I have seen it develop from the very beginning, before the first typed words hit the blogosphere; moreover, I personally know all the characters involved, particularly the chief villain in the story, the reason being that it is actually (in part) about me! On 21 February 2008

I met for coffee with an independent journalist, Suzan Mazur, who had called me a few days earlier saying that she worked for the *Philadelphia Inquirer* (turns out she had *written* for it, but was not a regular employee of the newspaper) and was working on a story on evolution and its critics. We met in the afternoon at Pret-a-Manger, on 42nd Street across from the New York Public Library, where I was doing some research for this book. The meeting was pleasant but also characterized by the usual mixed feelings a scientist has when talking to a journalist: on the one hand, I wanted to clearly explain why intelligent design is not science, on the other, I was aware that the journalist had only an approximate grasp of the matter under discussion and that very likely the article would come out as a combination of quasi-accurate statements and things that I had not actually said or meant. Still, I was in for quite a surprise when I read Mazur's account of the meeting a few days later.

The first of what turned out to be a series of articles by Mazur featuring me and several of my colleagues came out on 4 March, not in the *Philadelphia Inquirer*, but rather in a dubious online outlet based in New Zealand called *Scoop* (a publication that proudly announces that it does not edit anything that it publishes). The title of the article was "Altenberg! The Woodstock of Evolution?" To my astonishment, Mazur had essentially ignored everything I told her over almost an hour of chat and instead focused on the last five minutes of our conversation. Before leaving the coffeehouse, she had asked me what I was up to during the summer, and I casually mentioned some travel plans, including a workshop on the status of evolutionary theory I was organizing in Altenberg (Austria, near Vienna) with the logistical and financial support of the Konrad Lorenz Institute for Evolution and Cognition Research (KLI). The workshop was one of many hosted by the KLI, and one of hundreds organized yearly by various scientific organizations to get researchers to talk about whatever interests them. Since my technical work includes the conceptual and philosophical foundations of evolutionary theory, I had submitted a proposal to the KLI involving the invitation of sixteen well-known speakers in the field, with the intention (also not at all unusual) of eventually publishing an edited volume of the proceedings of the meeting.[12] But "the Woodstock of evolution"? Where did that come from?

From the fertile imagination of Mazur, that's where. She characterized the workshop as "a gathering of 16 biologists and philosophers of rock star stature," which she labeled "the Altenberg 16." The hype, rather disturbing to begin with, reached a crescendo throughout the article, where

Mazur stated that "despite the fact that organizers are downplaying the Altenberg meeting as a discussion about whether there should be a new theory, it already appears a done deal . . . Indeed, history may one day view today's 'Altenberg 16' as 19th century England's 'X Club' of 9—Thomas Huxley, Herbert Spencer, John Tyndall, et al.—who so shaped the science of their day." OK, few people would mind being compared to a rock star (though, alas, without the corresponding income), and any scientist would certainly enjoy being referred to as a modern day Thomas Huxley (he was the famous "Darwin's bulldog" of the Victorian age), but Mazur's presentation of the scope of the workshop was misleading at best. Yes, this was a gathering of well-known evolutionary biologists and philosophers, but the agenda was simply to discuss the current status of the theory and talk about what progress has been made in the past few years. Calls for an "extended evolutionary synthesis," that is, a modified version of the biological equivalent of the standard model in physics (the so-called Modern Synthesis), had been made before, and it was therefore not that unusual to get people together in the same room to talk about it. Things only got worse from there, and quite quickly too.

Mazur's long article turned out to be a hopelessly confused hodgepodge of actual science, badly misunderstood science, philosophy (good and bad), and crackpotism. For instance, one of the characters introduced by Mazur to the public was Stuart Pivar, an art collector who had been sending more or less threatening emails to me, my graduate students, and several of my colleagues because we were not taking seriously his "theory" about the evolution of development (Pivar's "theory" is based on the idea that all life is a variant of one basic form, a concept that arches back to pre-Darwinian times—and his "proof" is that he can build plastic models that can be reshaped as one wishes by twisting and turning them in a rough approximation of animal embryos). Mazur presented Pivar as a scientist who "is not dependent on government grants to carry out his work" (true, except that he is not a scientist, and I seriously doubt that he would be able to get government grants if he tried) and one who "has paid the price for [his unorthodox theory] on the blogosphere." In other words, Mazur painted me as a rock-star celebrity-rebel and Pivar as a misunderstood genius who is shunned by the scientific community because of a secret natural selection cabala. Needless to say, both portraits are as far from the truth as an episode of *The X-Files*. Mazur ended her puzzling article with a call for a "global" public broadcast of the Altenberg meeting, on the ground that a new theory of evolution would surely impact everybody's life.

This might have been nothing more than an amusing incident and one more reminder of what can happen when one talks to journalists without double checking their credentials. But we live in the era of the Internet, and the matter was far from being over. The very same day, Paul Nelson, a contributor to "Uncommon Descent," the official blog of pro-intelligent design author Bill Dembski, picked up on the *Scoop* article, declaring it worthy of attention on the grounds that it shows that "evolutionary theory is in—and has been, for a long time—a period of great upheaval. Much of this upheaval is masked by the noise and smoke of the ID debate, and by the steady public rhetoric of major science organizations, concerned to tamp down ID-connected dissent." Oh boy. To my further surprise, not only was Nelson's short commentary linked to by a variety of pro- and anti-evolution blogs, but eventually the Mazur article came to the attention of serious media outlets, including the *New York Times*, *Science,* and *Nature*. *Science* published a detailed piece the same week the workshop was being held, beginning with a whole paragraph devoted to Mazur's "coverage."[13] At least the *Science* reporter wryly pointed out that I am no Jimi Hendrix, switching to a more realistic portrayal of what we actually did in Austria that summer.

But Mazur herself wasn't done with it. She began publishing (also in *Scoop*) a six-part "investigative" story entitled "The Altenberg 16: Will the Real Theory of Evolution Please Stand Up?" Once again, this turned out to be a longwinded ramble that displayed in full color Mazur's lack of understanding of the subject matter and penchant for conspiracy theorizing. What was both illuminating and amusing was the shifting portrait of me that she presented to her readers. In the first article, back in March, I was introduced as that "rare combination—a consummate scientist with a sense of humor," and she made a charming reference to the fact that I had come to the interview carrying a birthday present in my hands (for my daughter). But subsequently I made the mistake of making clear—politely, I thought—that I was not too pleased with the distortions for which Mazur was responsible. This turned me into a rather shady character in her later articles. "Massimo Pigliucci is a man on the move," she wrote on 8 July, "but who is he? And why was he born in Liberia during the regime of William Tubman?" (because my father was there working for a British company that built roads, she could have found out, had she asked; besides, what does *that* have to do with anything?). From consummate scientist I had become "flamboyant" (something she clearly didn't mean as a compliment), and observations on my sense of humor had given way to remarks on my receding hairline (a true fact of life in my forties). Mazur, referring to a public talk I had given in New York and that she attended, wrote, “The second the event ended, Pigliucci fled. I took the opportunity to say hello to his wife, whose previous life as an adventurer and a director of the International Rescue Committee impressed me.” Just for the record: I stayed for a forty-five-minute question-and-answer session (which included answering one of Mazur’s own queries), followed by dinner with whoever wished to join us among those present at the event; oh, and my wife is not nor has ever been an “adventurer,” and she is not the director of the International Rescue Committee (she works for that humanitarian organization, in charge of their emergency-response unit). So much for accurate reporting in journalism.

I am happy to share that the “Altenberg 16” had a pleasant and intellectually stimulating meeting during the summer of 2008, though none of us felt like either Jimi Hendrix or Thomas Huxley. There were no smashed guitars during the presentations. The Mazur *Scoop* story is remarkable not for its consequences (it had virtually none), but because it allowed me a rare glimpse from the inside of the development of a journalist’s behavior once she thought (mistakenly) that she was onto something big. The fact that serious media outlets got interested in the story as soon as the Mazur article was out also makes one appreciate how thin the boundary is between not only science and pseudoscience, but journalism and pseudojournalism. What one should not, however, take away from this story is what many of my colleagues have too quickly concluded: that scientists should not talk to reporters. Scientists have an ethical duty to the public to explain what they do and why. This is because what we do may affect the public welfare, and even when it doesn’t, it is probably funded, directly or indirectly, by taxpayer’s money. Moreover, it simply does not help the image of science if scientists keep propagating the (somewhat founded) stereotype of the white coat aloof in the ivory tower. But by the same token, journalists also have a duty to do their homework and to present stories in an interesting, but not unduly sensationalist, fashion. We may all be after the Nobel or the Pulitzer, but few of us have even a fighting chance to get there. In the meantime, we ought to do our job, science or journalism, with the degree of seriousness that is rightly expected from professionals, not to mention quite simply the fact that we are all supposed to be grownups.

CHAPTER FIVE

Debates on Science

The Rise of Think Tanks and the Decline of Public Intellectuals

It is ironic that the United States should have been founded by intellectuals, for throughout most of our political history, the intellectual has been for the most part either an outsider, a servant or a scapegoat.—Richard Hofstadter

An intellectual is a person who has discovered something more interesting than sex.—Aldous Huxley

"Mr. President, Would you allow me, grateful as I am for the kind reception you once extended to me, to show my concern about maintaining your well-deserved prestige and to point out that your star which, until now, has shone so brightly, risks being dimmed by the most shameful and indelible of stains." So began a famous letter by the French writer Émile Zola to Felix Faure, then president of the French Republic, on 13 January 1898. Zola continued: "[I shall dare.] Dare to tell the truth, as I have pledged to tell it, in full, since the normal channels of justice have failed to do so. My duty is to speak out, not to become an accomplice in this travesty," and concluded the long letter with: "I have but one passion, the search for light, in the name of humanity which has suffered so much and is entitled to happiness. My fiery protest is simply the cry of my very soul. . . . I am waiting."

Zola's detailed and passionate appeal was prompted by the Dreyfus affair, a case that dominated the French public arena for several years between 1894, when artillery officer Alfred Dreyfus was suddenly arrested, and 1906, when he was officially exonerated of all charges. The episode was a complex story of international intrigue, military secrecy, political maneuvering, and a large dose of anti-Semitism, a cocktail that inspired countless newspaper editorials, several books, and of course a variety of movies (the first one, dated 1899, was produced in France while the controversy was still raging!). Zola's letter marked the opening salvo that eventually managed to get several prominent intellectuals involved in the public defense of Dreyfus, including the mathematician Henri Poincaré and the novelist Anatole France. Indeed, this episode is often considered the beginning of the modern figure of the "public intellectual," a crucial and yet much-maligned player in the "science wars" that concern themselves with the way science is presented and perceived at the onset of the twenty-first century—a subject we will take full on toward the end of the book.

In this chapter we will look at the alleged decline of the public intellectual, especially in the United States, as well as at the parallel ascent and evolution (some would say de-volution) of so-called think tanks. While any direct link between these two phenomena is difficult to establish, and at any rate it is a matter for historians of culture and sociologists to debate, I will treat both as rather disconcerting indicators of the level of public discourse in general, and of the conflict between science and pseudoscience in particular. It is a terrain fraught with complexities, where it is hard to make a clear distinction between heroes and villains, and where good ideas often result in questionable implementations. Nonetheless, it is an area that is both usually neglected within the context of discussing science in the public arena and yet crucial to our understanding of how science is perceived or misperceived by the public. Let us then begin with the modern intellectual heirs of Zola and company and see how they are managing little more than a century since the onset of public intellectualism.

Public Intellectuals in the Twenty-First Century: An Endangered Species or a Thriving New Breed?

Before we can sensibly ask whether public intellectuals are on the ascent, the decline, or something entirely different, we need to agree on what exactly, or even approximately, constitutes a public intellectual. It turns out that this isn't a simple task and that the picture one gets from the literature on intellectualism depends largely on what sort of people one counts as "public intellectuals," or, for that matter, what sort of activities count as intellectual to begin with. Nonetheless, some people (usually intellectuals) have actually spent a good deal of time thinking about such matters and have come up with some useful suggestions. For example, Amitai Etzioni, quotes the Enlightenment figure Marquis de Condorcet to the effect that intellectuals are people who devote themselves to "the tracking down of prejudices in the hiding places where priests, the schools, the government and all long-established institutions had gathered and protected them."[1] Or perhaps one could go with the view of influential intellectual Edward Said, who said that intellectuals should "question patriotic nationalism, corporate thinking, and a sense of class, racial or gender privilege."[2]

Should one feel less romantic (even a bit cynical, perhaps) about the whole idea, one might prefer instead Paul Johnson (also cited by Etzioni), who said that "a dozen people picked at random on the street are at least as likely to offer sensible views on moral and political matters as a cross-section of the intelligentsia."[3] Or go with David Carter, who wrote in the *Australian Humanities Review* that "public intellectuals might be defined as those who see a crisis where others see an event."[4] Or, again, we might be charmed by postmodernist Jean-Francois Lyotard, who said that a professor is "no more competent than memory bank networks in transmitting established knowledge."[5] And the list could go on and on, on both sides.

Regardless of how critical one is of the very idea of public intellectualism, everyone seems to agree that there are a few people out there who embody—for better or worse—what a public intellectual is supposed to be. By far the most often cited example is the controversial linguist and political activist Noam Chomsky. Indeed, his classic article "The Responsibility of Intellectuals," written in 1963 for the *New York Review of Books*, is a must-read by anyone interested in the topic, despite its specific focus on the Vietnam War (then again, some sections could have been written during the much more recent second Iraq War, almost without changing a word).

For Chomsky the basic idea is relatively clear: "Intellectuals are in a position to expose the lies of governments, to analyze actions according to their causes and motives and often hidden intentions. . . . It is the responsibility of intellectuals to speak the truth and to expose lies."[6] Yet one could argue that it is the responsibility of *any* citizen in an open society to do just the sort of things that Chomsky says intellectuals ought to do, and indeed I doubt Chomsky would disagree. But he claims that intellectuals are in a special position to do what he suggests. How so? It is not that Chomsky is claiming that only genetically distinct subspecies of human beings possess special reasoning powers allowing them to be particularly incisive critics of social and political issues. Rather, it is that intellectuals—at their best—are more insightful in social criticism because they can afford to devote a lot of time to reading and discussing ideas—something that most people trying to make a living simply do not have the time or energy to do. Moreover, intellectuals have a *duty* to be so engaged with society because often their vantage point is the result of a privileged position granted them by society, most obviously in the case of academic intellectuals (but also journalists, some artists, and assorted others), who are somewhat shielded by most direct political influence or financial constraints, and whose professional ethos requires intellectual honesty and rigor.

Of course, none of this guarantees that public intellectuals always get it right, or that they always further the welfare of society. The classic counterexample, mentioned by Chomsky himself, is the philosopher Martin Heidegger. Heidegger is a controversial figure, to say the least, both academically and politically (not unlike Chomsky himself, though the similarity ends there). Some commentators consider him one of the greatest of modern philosophers; others think that his writings are full of obfuscatory language and sheer nonsense. He was the mentor of Leo Strauss—who in turn inspired the modern neoconservative movement—as well as the father of several movements that feature prominently in the "culture wars," such as deconstructionism and postmodernism. At any rate, Heidegger was elected rector of the University of Freiburg in Germany in 1933, under the auspices of Hitler's regime. The inaugural address he delivered is in fact a good example of convoluted nonsense, but it just as clearly represents the exact antithesis of what a public intellectual should be. For example, after having waxed poetic about spiritual missions and the "essence" of German universities, Heidegger went on to say that "German students are on the march. And whom they are seeking are those leaders through whom they want to elevate their own purpose so that it becomes a grounded, knowing truth," a rather ominous presage of things to come for the German youth. And he kept going: "Out of the resoluteness of the German students to stand their ground while German destiny is in its most extreme distress comes a will to the essence of the university . . . The much-lauded 'academic freedom' will be expelled from the German university."[7] Heidegger's connection with the Nazi will forever taint his legacy, making him a permanent warning to aspiring public intellectuals about what route not to follow.

Chomsky, on the other hand—in a prose infinitely clearer and more compelling than Heidegger's—raises the question of what the duty of an intellectual ought to be, and answers that she has to be concerned with the creation and analysis of ideologies, including those endorsed or produced by intellectuals themselves. Since the public intellectual, according to Chomsky, has to insist on truth, she also has to see things in their historical perspective, truly to learn from history rather than be bound to repeat the same mistakes over and over. Of course, as we shall see later in the book, the whole concept of "truth," and by implication the efficacy of both science and of intellectual discourse, has been questioned by those academic heirs of Heidegger known as postmodernists. Setting that aside for a moment, however, there are in fact other ways of being skeptical of the whole idea of public intellectualism, for example in the analysis of Richard Posner, author of *Public Intellectuals: A Study in Decline*.

Posner was a professor at the University of Chicago Law School and later became a judge on the U.S. Court of Appeals (Seventh Circuit), where he was nominated by President Reagan. By all accounts, Posner is considered a major and influential legal theorist. Despite being a highly regarded intellectual, his analysis of intellectuals as a breed is anything but sympathetic, although he makes several interesting points that we need to consider while attempting to put together a general picture of intellectualism and how it relates to public understanding of science. Posner, by his own account, wrote his book as a result of what he perceived as the low quality of public intellectuals' commentaries in two high-profile cases in which he was involved: the impeachment hearings of President Clinton and the antitrust case against Microsoft. Posner's thesis is that there has indeed been a decline of intellectuals in the United States, but that this isn't a matter of fewer of them being around. On the contrary, the "market" for intellectualism has allegedly increased dramatically in recent years, but the quality of the individuals populating such a market has decreased sharply.

Posner accounts for this double trend (increase in quantity and decrease in quality) with an ingenious, if certainly debatable, analysis of some of the forces shaping both the supply of and the demand for public intellectuals. On the supply side, intellectuals are now almost exclusively an academic phenomenon. Gone are the days of Zola and (Anatole) France, when it was the independent artist or writer who was more likely to be outspoken about social matters. Instead, the rise of universities (in terms of both numbers and financial resources) after World War II has catalyzed a shift toward academic-type intellectuals. Academics are better positioned than independents to play a role in public discourse because they are more readily perceived as credentialed individuals (after all, most of them sport Ph.D.s and often work at prestigious institutions, such as Posner's own University of Chicago), and they can afford (literally, in financial terms) to stick their neck out about controversial matters in the relative safety of a tenured position—although the tenure system itself is increasingly under attack, partly for ideological reasons related to the discomfort often created by academic intellectuals.

Not that all is good and well for academics who wish to venture in the public arena. As Posner points out, there are pros and cons that need to be carefully evaluated. On the side of incentives there is the possibility of monetary reward (academic salaries aren't what they used to be), though it is fairly rare that an academic actually lands a major book contract or is in sufficient demand to command significant honoraria for speaking engagements.

On the side of disincentives, there are several, some potentially career-crippling. To begin with, the more time an academic devotes to speaking and writing for the public, the less she has to engage in scholarship and research—and it is the latter that gets you tenure and promotions, which helps explain why most public academics are middle aged, post-tenure, and possibly past their intellectual prime. Moreover, the myth of the ivory tower is anything but a myth: despite occasional protestations to the contrary, most academics themselves see engaging the public as a somewhat inferior activity, sought after by people who are vain, in search of money, not particularly brilliant scholars, or all of the above. According to Etzioni in his coedited book mentioned above, after his death astronomer and science writer Carl Sagan (to whose career we will turn in some detail later in this chapter) was referred to as a "cunning careerist" and a "compulsive popularizer"—not exactly encouraging words for other scientists considering following in his footsteps.[8]

Posner also turns his analysis to the other side of the coin, looking into what sort of demand there might be for public intellectuals. Before I discuss his viewpoint, I must say upfront that I think that "market analyses" of this issue, although relevant to societal dynamics, miss the major point: regardless of what sort of market there might or might not be for public intellectuals, I argue that it is still a moral duty of all citizens, and of intellectuals in particular, to intervene in public discourse. In this I am squarely on Chomsky's side. Be that as it may, Posner's points are still well taken.

He suggests that there are at least three "goods" that public intellectuals may be "selling" (sometimes literally, although as I mentioned earlier, much work done by intellectuals in this respect is actually pro bono), and that therefore influence the level of "demand" for intellectuals themselves. Besides the obvious one, that is, presumably authoritative opinions on current issues of general relevance, there are what Posner calls "entertainment" and "solidarity" values.

There is little question that we live in a society in which entertainment, broadly defined, reigns supreme. The nightly news, not to mention the twenty-four-hour news channels, are increasingly less about serious journalism and more about sensationalism or soft news, so much so that one can make a not entirely preposterous argument that *The Daily Show* with Jon Stewart is actually significantly more informative than the real news shows that it is meant to spoof. As biologist Richard Dawkins lamented,[9] we think that our kids need to have "fun, fun, fun" rather than, say, experience wonder or interest (they are not the same thing) when going to school or a museum. And social commentator Neil Postman has convincingly argued that our society is about *Amusing Ourselves to Death*, as he put it in the title of one of his most provocative books. It is therefore no surprise that even intellectuals have to possess an entertainment value of sorts, although it is of course difficult to quantify it and its effect, if any, on the demand for public intellectuals. Such effect will also depend greatly on the specific media outlet: while there are plenty of TV channels and newspapers for which the entertainment value is probably close to the top of priorities, there are still serious media outlets out there (BBC radio and TV, National Public Radio, Public Television, the *New York Times*, the *Washington Post*, the *Guardian*, the *Economist*, Slate.com, and Salon.com come to mind, among many others) where the relevance and insight of what the intellectual has to say are paramount.

We come next to the concept of "solidarity value" proposed by Posner. This is often underestimated, but I suspect it does play an increasingly important (and, unfortunately, negative) role in public discourse. The idea is that many, perhaps most, people don't actually want to be informed, and even less so challenged in their beliefs and worldview. Rather, they want to see a champion defending their preconceived view of the world, a sort of ideological knight in shining armor. Blatantly partisan outlets such as Fox News (on the right), Air America (on the left), and the countless number of Evangelical Christian radio stations are obvious examples of this phenomenon, but perhaps the most subtle and pernicious incarnations of it are all over the Internet. The characteristics of that medium are such that it is exceedingly easy to customize your access so that you will only read what people "on your side" are saying, never to be exposed to a single dissenting viewpoint. While blogs, for example, are indubitably a revolutionary and potentially very powerful way to expand social discourse, it is also very easy to bookmark or subscribe by feed to a subset selected in order to further entrench, rather than challenge, your opinions. (That said, my experience with my own blog, rationallyspeaking.org, is one of animated debate among people with diverging perspectives, including readers who alternately agree or disagree with my posts, depending on the context. Seems like the blogosphere is ripe for some quantitative social science analysis.)

All things considered, Posner's arguments point toward a level of supply and demand for public intellectuals that translates into a larger number of them than probably at any time in history. But, of course, quantity is rarely an indication of quality, which brings us to Posner's contention that the decline of the intellectual is a matter of lowered quality, not of insufficient numbers. There are fundamentally three reasons for this conclusion, all of which are circumstantial, as it is very hard to assess the quality of public intellectual discourse in any objective and statistically quantifiable way. We have already seen the first reason: since intellectuals are sought after at least in part for their entertainment and solidarity values, and given that neither of these is presumably related at all to the degree of insight offered by the opinions being delivered, quality is liable to suffer.

The second reason advanced by Posner to conclude that public intellectuals just aren't what they used to be is the inevitable march of academia toward increased specialization of its scholars. Remember that most modern intellectuals are academics, and they are successful within academia because they specialize on incredibly narrow fields of scholarship and research—since most of the broad ones have already been covered exhaustively by their predecessors. For example, the joke used to be that philosophers are people who know nothing about everything (they are intellectual generalists), while scientists know everything about nothing (they are intellectual specialists). But in today's universities, even philosophers are converging toward the stereotype of the scientist: I know colleagues in philosophy departments who spend a lifetime publishing analyses of the work of just one (usually long-dead) philosopher, and often not a major one at that. Similarly, some of my colleagues in science think it absolutely crucial to invest many hundreds of thousands of dollars and countless graduate-student years to figure out whether an obscure species of mushroom is by any chance found also in Antarctica. Don't get me wrong: this is the way it must work if one wishes to contribute something truly novel to one's field at the dawn of the twenty-first century. But such an astoundingly narrow focus hardly makes for a Zola-type figure: if someone has to spend most of his productive life reading and working on very specific problems, how is the same person then supposed to suddenly turn around and knowledgeably pontificate about the big issues concerning our increasingly complex society? (Incidentally, as pointed out by Murray Sperber in his *Beer and Circus: How Big-Time College Sports Is Crippling Undergraduate Education*, this is also what explains why the idea that an excellent scholar makes for an excellent teacher is, in fact, a myth propagated by both academics and administrators for exclusively self-serving reasons.)

Finally, there is the failure of the so-called marketplace of ideas, which Posner is one of the few to keenly recognize. The phrase originated with Justice Oliver Wendell Holmes, who wrote the dissenting opinion in the infamous *Abrams v. United States* case argued in 1919 in front of the Supreme Court. The case was a test of a law passed the year before, which made criticism of the U.S. government a criminal offense. The law was upheld, and the statute not invalidated until *Brandenburg v. Ohio*, during the Vietnam War. Holmes wrote passionately about the safeguard for freedom of speech enshrined in the American Constitution, arguing that "the ultimate good desired is better reached by free trade in ideas—that the best test of truth is the power of the thought to get itself accepted in the competition of the market, and that truth is the only ground upon which [men's] wishes safely can be carried out. That at any rate is the theory of our Constitution." Holmes was correct in suggesting that a necessary condition for maximizing our chances to find the truth about whatever subject matter or for reaching a consensus on moral and social issues is to allow ideas to "compete" for people's minds and hearts. But Posner's well-taken point is that this is by no means a *sufficient* condition.

There is need of a second factor in addition to the free marketplace of ideas, and this second factor, according to Posner, is sorely lacking. You see, in order for the best ideas to win the competition the judges must be, well, competent. But the judges here aren't indisputable facts that can be verified by anyone; they are the opinions of a generally badly informed and undereducated (with respect to the relevant issues) public. This is a public that has little time for the sort of in-depth analyses and research that would allow it to actually assess the contributions of intellectuals on their merit. Ironically, this is precisely why Chomsky says that it is up to intellectuals, not the public, to do the hard work of research and documentation. The problem, Posner suggests, is that the public tends to go by much less reliable proxies of quality, such as credentials (but, really, it isn't that difficult to get a Ph.D., even from a reputable university) and the rhetorical abilities of the intellectuals themselves. Of course, an intellectual that makes a mark by virtue of his reputation and rhetoric, rather than because of the content of his ideas, is the quintessential sophist so harshly criticized by Socrates in several Platonic dialogues twenty-four centuries ago. Sometimes there really hardly seems to be anything new under the sun.

There is much to be commended in Posner's analysis of the decline of the modern intellectual, and yet one cannot help thinking that it is somewhat self-destructive for intellectuals such as Posner to be so critical of their own role and dynamics in society. A good reality check is a positive and necessary part of public as well as academic discourse, but an exceedingly negative attitude soon breeds contempt for intellectual discourse itself and a nihilistic dismissal of it.

A more positive approach to the analysis of contemporary intellectualism is perhaps provided by Frank Furedi in his *Where Have All the Intellectuals Gone?* which starts out not with self-criticism, but with ridiculing politicians in charge of public education, namely, the then Labour secretary of state for education in the UK, Charles Clarke. Clarke characterized the idea of education for its own sake—the very foundation of the so-called liberal educational approach common in modern universities—as "dodgy," which can be interpreted to mean anything from dishonest and unreliable to potentially dangerous. Clarke's opinion is that the government should not support "the medieval concept of a community of scholars seeking truth."[10] Along similar lines, former U.S. President Ronald Reagan infamously said during his campaign for governor of California in the late 1960s that universities should not subsidize intellectual curiosity.

This sort of nonsense does "trickle down" (a phrase surely Reagan would have appreciated) to university communities. For example, several years ago a librarian friend of mine was disgusted when the president of the community college where she was working in the United States gave a "pep talk" to the staff at the beginning of a new academic year. Instead of waxing poetic about the empowering effects of a solid education, the administrator in question bluntly compared the students to "customers" whose needs have to be satisfied at all costs. "Imagine yourself selling them ice cream." (I slightly paraphrase from memory here.) "If they want pralines on top, you sprinkle pralines on top." No doubt an uplifting vision for the educators gathered there to listen to this abominable, but increasingly common commodification of higher education.

Of course Furedi is fully cognizant of the fact that a liberal education, just like the idea of a fearless public intellectual, is an ideal and has never corresponded to a historical reality at any point in the past. But it makes a big difference in terms of attitudes and motivations whether we consider an ideal a goal to strive for or we dismiss it as irrelevant, outdated, or even positively dangerous.

While Furedi devotes some space to blaming postmodernism for the decline of intellectualism, a more intriguing observation is that intellectualism is on the retreat just at the time we keep hearing of a "knowledge economy" and when bookstores, book clubs, poetry readings, museums, and galleries are doing increasingly well, to the point of having a single TV show in the United States (*The Oprah Winfrey Show*) determining which book will become the next overnight bestseller. Yet this apparent paradox is actually explainable by the same distinction that Posner made about intellectuals themselves: quantity does not necessarily translate into quality. While bookstores are increasingly popular, there are basically only two or three major chains left in most parts of the United States (and at least one of them is rumored to be close to bankruptcy), which means that a small number of individuals wields a huge amount of decision power when it comes to which books to promote and which to relegate to the back shelves or even keep out of the store (and, largely, off the market). Winfrey, to her credit, has almost singlehandedly made the rather esoteric and somewhat snobbish idea of a book club one of the most popular activities engaged in by scores of Americans. But then again, some of her picks have been embarrassing, as in the infamous case of James Frey (author of *A Million Little Pieces*), who turned out to have made up large parts of his allegedly autobiographical story (Winfrey eventually challenged him on her show, but her initial response was that he may have been telling a subjective truth—a perfect postmodernist and nonsensical way of saving face).

Furedi identifies another culprit in the ongoing quest for the disappearance of the public intellectual: the assault on meritocracy. Americans in particular have always had a rather bipolar attitude toward meritocracy: on the one hand, the United States was established by people whose very creed included the idea that it is merit, not birthright, that ought to determine one's fortunes. American bookstores abound with large sections of books written by successful people telling others how to become successful (here is one hint: write a book about how to become successful), and American CEOs and sports figures are lionized because of their *merits*, not because of their family trees. Then again, one of the reasons Al Gore allegedly lost the 2000 presidential election against George W. Bush is because Gore was seen as an "egg-headed intellectual," obviously an insult, not a compliment.

Furedi's take is that over the past several decades the very conception of meritocracy has shifted from a powerful incentive paving the way to a fairer society to an intrinsically anti-egalitarian and undemocratic tool of oppression. The reasons for this are many and complex, and they include the rise of new philosophies of teaching within education departments at colleges throughout the nation, as well as the recognition of the sociological fact that certain minorities tend to be at a disadvantage in our meritocratic system as presently constituted (this is not, obviously, a point in favor of biological racism, but a concession to the real difficulties in overcoming cultural conditions and in creating a true level playing field). Indeed, Furedi proposes the bold thesis that the currently entrenched rejection of meritocracy is based on two mistaken ideas: first, that a large section of the population is somehow intrinsically incapable of achieving high academic standards; and second, that this failure leads to mental distress, and therefore it is appropriate to substitute "feeling good" for actual results. But this is a recipe for disaster, because, as Furedi puts it, "Rewarding merit implies treating people as adults, whereas magicking away the sense of failure is motivated by the desire to treat them as children."

The Many Faces of Anti-Intellectualism

No discussion of the sort we are engaged in here could be complete without a mention of Richard Hofstadter's *Anti-Intellectualism in American Life*, which appeared in 1963. In *Denying Evolution: Creationism, Scientism and the Nature of Science*, I stressed the point that to understand the peculiar (within the West) phenomenon of creationism one has to put it into the broader context of the general history of anti-intellectualism in the United States. This is a fortiori true when the scope of our analysis is the whole of the scientific enterprise and how it is viewed by the media and the public, as in this book.

The starting point of Hofstadter's essay is, naturally enough, to produce some sort of definition of anti-intellectualism. It would be unproductive to attempt precise definitions, based on sets of logically necessary and sufficient conditions, for any complex concept. This is social science, not mathematics. Nonetheless, meaningful discourse is possible only if all interested parties have a reasonably sharp idea of what it is that the discourse is about. Hofstadter himself puts it this way: "Anti-intellectualism is not a single proposition but a complex of related propositions. . . . The common strain that binds together the attitudes and ideas which I call anti-intellectual is a resentment and suspicion of the life of the mind and of those who are considered to represent it; and a disposition constantly to minimize the value of that life." Hofstadter (and a later commentator, Daniel Rigney[11]) goes on to identify three fundamental kinds of anti-intellectualism, tracing their origins and exposing their characteristics: anti-elitism, instrumentalism, and anti-rationalism.

I have briefly touched on the paradox of anti-elitism in American society when discussing the work of Furedi, but it is important to realize that the Gore-Bush contrast is just one example in a long series. Thomas Jefferson was the first presidential candidate to be criticized on the grounds of his "theoretical" leaning and association with Enlightenment France, and this was a mere twenty-five years after the Declaration of Independence! But the turning point in American political history as far as this discussion is concerned was arguably the 1828 election, in which Andrew Jackson beat John Quincy Adams. Adams was a Jefferson-type intellectual, member of the cultural (and economic) elite that had led the country during the War of Independence and through the forging of the Constitution and the Bill of Rights. Jackson, by contrast, was a man of action, "unencumbered by the weight of too much book learning," as Rigney puts it. Ever since Adams's defeat, intellectuals have had a hard time getting elected presidents, while populists (sincere or not) have had a field day. (The recent and spectacular election of Barack Obama was an exception; another one was Theodore Roosevelt, on whom, as Rigney remarks, it was hard to pin the common stereotype of intellectuals as effeminate or unmanly—never mind the remarkable fact that such attributes are still considered by far too many people as an insult.) Historically, it was George Wallace, as late as the 1960s, that actually coined the disdainful phrase "pointy-headed intellectuals," later recycled in modified form on the occasion of the Gore-Bush rivalry.

As I mentioned earlier, anti-elitism is complex and contradictory within American society, as this is a culture where elitism stemming from birth or intellect is seen with deep suspicion, and yet elitisms based on money (itself highly culturally heritable) or physical prowess are elevated to mythical levels. Americans have not yet accepted as a culture that the fact that people are born with different abilities (be they mental or physical) does not translate into the value judgment that they therefore ought to be accorded different status as human beings. Until they do, anti-elitism will in fact seriously limit the impact that intellectuals might have on the political and social discourse in the United States.

Instrumentalism is another quintessential American ideology, although it is not an all-out anti-intellect position. The idea is that learning is to be encouraged if it is aimed at addressing practical concerns (after all, even within the highly esoteric field of philosophy, Americans are most famous for the school of *pragmatism* embodied by John Dewey and William James, among others). The roots of instrumentalism can be found in the influence of the "robber barons" (or, depending on one's social and political position, "captains of industry") that provided the foundations for American capitalistic society itself. Rigney provides us with a series of quotes that make the point very clearly. For example, Andrew Carnegie famously said that classical studies are a waste of "precious years trying to extract education from an ignorant past." The fact that Carnegie is now remembered for having endowed a university by the same name in Pittsburgh, for having donated large amounts of money to public libraries, as well as for the world-famous Carnegie Hall in New York City (where a very classical sort of art is performed almost daily) is yet another deliciously ironic contradiction in this debate (it is to be noted, incidentally, that Carnegie, Vanderbilt, and others were of course the quintessential incarnation of a money-powered elite).

Instrumentalism makes less and less sense the more a society develops above the level of direct struggle for life. At the beginning of U.S. history, just as in many parts of the world still today—unfortunately—the imperative was first to survive from one year to another, or even one day to another, and then to lay the foundations for a more stable and prosperous society. But as soon as basic human needs are met and a minimum sense of security settles in, we begin to look for other ways to fulfill our lives, what Aristotle referred to as "flourishing." A liberal education is not meant to provide people with the know-how for functioning in either an agricultural or an industrial setting; its goal is to produce mentally sophisticated citizens, capable of critical thinking, who can be fulfilled members of a complex society in which labor is a means to many ends, not just to survival. Of course, one could easily make the argument that "captains of industry" (and a good share of the politicians whose election campaigns are lavishly funded by them) have a vested interest in keeping the focus on teaching skills that are immediately useful to their enterprises and in discouraging any education beyond that. As Neil Postman again phrased it in the title of one of his books, education can be a highly subversive activity indeed. Within the context of our discussion, then, instrumentalism is one of the reasons why liberal education has been under constant attack from conservative politicians for decades, and it is why the National Science Foundation (which has a federal mandate to fund both basic—that is, not applied—scientific research as well as science education) tends to be among the least-funded science agencies in the country.

Perhaps the most complex, and pervasive, form of anti-intellectualism is the last one we will discuss here: anti-rationalism. This one has profound religious roots, tracing back at least to the first "Great Awakening" that led to the creation of the evangelical movement in the United States during the early 1700s. Ironically, the Awakening was a revolt against what was perceived as the hyperintellectual flavor of early American Puritanism, characterized by ministers who spoke of abstruse theological issues in highly philosophical terms rather than addressing the emotional concerns of their very flesh-and-bone followers in the pews. It is not difficult to see how modern-day evangelical and charismatic movements are direct cultural descendants of that first revolt. This in turn has led to the rise of the extreme religious right in politics and to the never-ending battle between evolution and creationism in science education.

Hofstadter discerned two major "currents" within anti-rationalism. In one current there is a tendency to see reason as cold, opposed to emotions, and therefore almost inhuman. One can think of this as the prototypical anti-Aristotelian position, since for the Greek philosopher reason is the highest characteristic of humanity and ought in fact to keep emotions in check (though he would have also immediately added that this is a matter of balancing the two, not simply of repressing emotions). The other current pits absolutism against relativism, particularly when it comes to moral problems. The fear here is that if, as postmodernist Michel Foucault put it, there is "no truly universal truth," then how do we tell right from wrong and therefore keep society from a rapid descent into anarchy and self-annihilation? Of course, one doesn't have to buy into Foucault's exaggeration to keep a skeptical view of absolute truth. It is a false dichotomy to think that there are only two positions: either we can have certain knowledge or anybody's opinion is as good as anybody else's. But it is this fear that demonstrably fuels, for example, the debate about evolutionary theory: after all, it isn't that people are particularly interested in the vagaries of the history of life as manifested in the characteristics of obscure species of insects or plants. The fear that strikes the creationist's heart when he contemplates what philosopher Daniel Dennett called "Darwin's dangerous idea" is that the modern scientific account of the origin of humans and of all other species directly contradicts the accounts in Genesis. And if Genesis turns out to be wrong about factual matters, how can we trust it when it comes to questions of morality?

Considering, as we have so far, the decline in quality if not quantity of intellectualism in the United States, as well as developing a better appreciation of just how deep the roots of anti-intellectualism are in American culture and history, should make it clear why the task for scientists who wish to do their part as public intellectuals seems daunting indeed. However, a few shining lights have emerged over the years, and we shall take a brief look at the career of two of these figures as an example of the perils of "going public" when one is a practicing scientist.

Carl Sagan and Stephen Gould: Case Studies in Scientific Public Intellectualism

I never got to meet either Carl Sagan or Stephen Gould, but they both had a profound effect (not always positive) on my career choices, the way I look at science, and even the very decision of writing this book. I encountered Sagan first, when my grandfather gave me a book Sagan wrote, *Broca's Brain: Reflections on the Romance of Science*, back when I was in junior high school. To this day, one of the most inspiring books of science for the public I've ever read is Sagan's *The Demon-Haunted World: Science as a Candle in the Dark*. I discovered Gould a bit later, when I was finishing high school and started reading his collection of essays from *Natural History*, beginning with *Ever Since Darwin*. His last technical book, *The Structure of Evolutionary Theory*, is one of the most irritating pieces of scientific literature ever written (he leaves page-long footnotes for the benefit of future historians of science interested in his thought processes) and yet is also full of valuable ideas that will keep biologists thinking for a long time to come. By any definition of a scientist as public intellectual, both Sagan and Gould were definitely among the most publicly intellectual of modern scientists, and they make good case studies because their lives and careers are well known and documented, because they engaged in some of the most controversial science issues when it came to public debate (Sagan wrote on both nuclear disarmament and global warming, Gould expounded on creationism and the relationship between science and religion) and because they both were severely criticized, indeed sneered at, by some of their less-famous colleagues, accused of being careerists with little to contribute to science.

Carl Sagan was probably most famous for his *Cosmos* television series, the most successful series ever aired by the American Public Broadcasting Service (PBS). He was born in Brooklyn in 1934 into a Jewish family, and while by all accounts he retained a strong attachment to Jewish culture, he was nonreligious. Scientifically, his career is deeply entwined with both planetary exploration (he worked at NASA's Jet Propulsion Laboratory and was involved in space missions from the Venus-bound *Mariner* 2 to the *Pioneer* and *Voyager* flights to the outer solar system) and the search for extraterrestrial intelligence (Sagan was instrumental in sending several messages out to potential ETIs, from the radio signal emitted at Arecibo in 1974 to the plaque on the *Pioneer* spacecraft to the golden disk with sounds from Earth aboard *Voyager*).

Michael Shermer, in *The Borderlands of Science*, makes the intriguing suggestion that SETI is the highbrow equivalent of UFOs, that the first is elitist and the latter is populist—a fitting contrast in the context of our discussion on anti-intellectualism. The implication is not that SETI is as much of a crackpot activity as ufology, but that the latter has always been dismissed by scientists while the former made a remarkable shift from fringe science to accepted orthodoxy. Largely this shift was the result of Sagan's advocacy (not of new discoveries, since, as we noted earlier, we still have no evidence of any ETIs anywhere in the universe). Even more interestingly, Sagan actually wrote on both SETI and UFOs, though from very different perspectives. As early as 1966 Sagan was a member of the committee reviewing the so-called Project Blue Book, the U.S. Air Force task force dealing with UFO claims. The report of the committee concluded that Blue Book was not a proper scientific study and that research on the topic should be shifted to a university setting. Three years later, Sagan pushed for and obtained the approval of the American Association for the Advancement of Science to hold a symposium on UFOs, featuring skeptics and supporters alike, and he edited the resulting proceedings in a 1972 book, *UFOs: A Scientific Debate*. This annoyed some of his colleagues to no end, a particularly unwise move for a young scientist (and the beginning of a pattern of behavior that doesn't fit too well with others' accusations that he was a careerist).

It is important to realize that Sagan was not a supporter of the ETI interpretation of UFOs: he did not think there was much evidence to back the sensationalist claims circulated by the press, and he thought of UFOs more like a quasi-religious phenomenon based on potent emotional responses. Conversely, Sagan thought of SETI as a scientifically credible enterprise, although of course he was as passionate about it as any ufologist is about flying saucers.[12] But when Sagan began talking about the search for extraterrestrial intelligence, the issue was not far from the level of ufology in the eyes of most practicing scientists, and it took decades of effort and advocacy for the situation to change. There is no question that Sagan became famous partly because of such advocacy—including the "stunts" he managed to pull off by convincing NASA to send messages to perspective ETIs—but it was a bold move that had little chance of success in the beginning, and it demonstrates the intellectual courage that characterized Sagan throughout his career.

Sagan committed some blunders, such as his prediction of a major environmental catastrophe as a result of the bombing of the oil fields in Kuwait during the first Gulf War. He was harshly criticized by colleagues, but he also admitted such mistakes and used them as a pedagogical tool to teach about the provisional nature of science in *The Demon-Haunted World*. Criticism by his peers often hit below the belt, as when physicist Edward Teller (who has himself been likened to the fictional character of Dr. Strangelove because of his unwavering support for the use of nuclear weapons) said to one of Sagan's biographers: "Who was Carl Sagan? He was a nobody! He never did *anything* worthwhile. I shouldn't talk with you. You waste your time writing a book about a nobody." Despite such a vituperative comment, Michael Shermer decided to investigate in a quantifiable manner whether Sagan had in fact done nothing worthwhile, at least when compared to some of his most esteemed colleagues.

Sagan was famously rejected by the American National Academy of Science (to which his first wife, Lynn Margulis, was admitted). Shermer wanted to compare the scientific c.v.'s of members of the academy to Sagan's own to see whether the latter would stand up to the group that rejected him. Alas, such information is not readily available from the Academy, so Shermer instead compared Sagan's output to that of four other eminent scientists of his time (all biologists):[13] Ernst Mayr, one of the coauthors of the Modern Synthesis, the currently accepted version of evolutionary theory; Jared Diamond, author (among other things) of *Guns, Germs and Steel*, which presents a theory of why some civilizations have flourished and others have struggled (we encountered him before when talking about history as a science); Edward O. Wilson, the father of the controversial fields of sociobiology and evolutionary psychology; and the very same Stephen Jay Gould that we shall discuss next. The results are quite clear: Sagan wrote far more books than all the others (most of them nontechnical, about popular science), he was fourth (ahead of Wilson) in terms of total numbers of scientific papers, and third (after Gould and Diamond) in the number of scientific papers per year. In other words, he was as technically prolific as any of the four excellent scientists to whom Shermer compared him, and he outdistanced them all when it came to book publishing. Of course, any practicing scientist will testify that there is no way that Sagan could have actually personally done all the research that went into maintaining an output of more than a paper a month for fourteen years (from 1983 until his death in 1996), but that is because the nature of modern science is collaborative, so that colleagues and graduate students (especially graduate students) share the burden. Accordingly, Shermer's data show that Sagan had by far the highest number of coedited and coauthored works of the group of five.

Carl Sagan was not a saint, nor should that be part of the expectation for a public intellectual. His first son, Dorion, accused him of "champagne socialism" for expounding the virtues of communism while dining at the Ritz Carlton, and first wife, Lynn Margulis (Dorion's mother), said of him that "he never changed a diaper in his life, he never cleared the table of his dishes, he never washed the dishes. . . . He needed ten thousand people to be raving about him all the time." Then again, his third wife, Ann Druyan, had a different take:[14] "If it was careerism that motivated him, surely he wouldn't have turned down three dinner invitations to the Reagan White House. . . . No careerist would have resigned from the Air Force Science Advisory Board and surrendered his top security clearance in protest over the Vietnam War." And, of course, the only reason we know all of this about Sagan is precisely because of his high-profile career in the public eye. I doubt that the average member of the National Academy of Sciences can claim a significantly higher ground in public matters.

Stephen Jay Gould grew up in another Jewish family, not far from Sagan's, in the neighboring New York City borough of Queens. Also like Sagan, he got his first academic position at Harvard, and—unlike Sagan (who spent most of his career at Cornell University)—he stayed there his whole life. Also unlike Sagan, Gould was actually admitted into the National Academy of Sciences in 1989, and his admittance as a fellow of the American Association for the Advancement of Science in 1983 explicitly cites his "numerous contributions to both scientific progress and the public understanding of science," a clear reference to his public-outreach work.

Part of the difference in the trajectories of the two scientists is no doubt due to their distinct styles and scientific contributions, but it is hard to escape the feeling that the generally more positive reception of Gould's public career is a (most welcome) sign of the times: after decades of ignoring or dismissing threats from creationism and other pseudoscientific attacks, the National Academy, the AAAS, the American Institute of Biological Sciences, and individual scientific societies such as the Society for the Study of Evolution have devoted an increasing amount of resources to public outreach. Indeed, the Society for the Study of Evolution has recently instituted a "Gould Prize" to recognize scientists who distinguish themselves in the public arena.

This doesn't mean that Gould's career was unaffected by his dual role as a scientist and a popularizer. John Maynard Smith, a prominent evolutionary biologist and critic of Gould, for example, was quoted in the *New Yorker* as saying that "Gould occupies a rather curious position, particularly on his side of the Atlantic. Because of the excellence of his essays, he has come to be seen by non-biologists as the preeminent evolutionary theorist. In contrast, the evolutionary biologists with whom I have discussed his work tend to see him as a man whose ideas are so confused as to be hardly worth bothering with, but as one who should not be publicly criticized because he is at least on our side against the creationists."[15] This is a rare and stunningly vitriolic assault on a colleague, which manages at the same time to dismiss the other's scientific accomplishments while making clear that his public work, while valuable, is certainly insufficient to redeem him professionally. Not that Gould himself didn't share in the blame: the same article, by Robert Wright,[16] referred to him as "the accidental creationist," because Gould was so hyperbolic in his criticism of standard evolutionary theory (arguably, in order to highlight the novelty of his own work) that creationists often quoted him as being on their side (he wasn't).

Gould's main contribution to evolutionary biology was the idea of "punctuated equilibria," which he proposed in 1972 together with fellow paleontologist Niles Eldredge. Punctuated equilibria essentially argues that the incompleteness of the fossil record, which so bothered Darwin, is not solely due to the fact that fossilization is a rare phenomenon. Instead, the record should be read as truly reflecting long periods of time during which organisms do not appear to change (stasis) punctuated by relatively sudden (in geological terms) spikes of morphological change. This tension between stasis and fast change, according to Eldredge and Gould, is a real feature of evolution, not an artifact of missing data, and ought to be considered in expanding the classical theory of evolution—which has often been described instead as a theory of gradual change.

The idea of punctuated equilibria has had the verifiable effect of producing a large scientific literature debating it, including the publication of countless new empirical papers. It has also brought paleontology back to the center stage of evolutionary studies and has forced scientists to seriously reconsider concepts such as species-level selection and the role of constraints imposed by developmental processes (which, according to Gould, are primarily responsible for stasis) in evolution. By these criteria alone, Gould (with Eldredge) deserves to be thought of as one of the most influential biologists of the twentieth century, and his contribution can hardly be dismissed as cavalierly as Maynard Smith did.

What of Gould as a public intellectual? In that arena, he started out like Carl Sagan, in a rather uncontroversial fashion, by writing essays explaining science to the general public. But, again like Sagan, Gould then moved to more contentious areas, crossing the fuzzy line from science popularizer to public intellectual. While Sagan took on global warming and nuclear war, Gould concentrated on racism and the relationship between science and religion. Gould, together with his colleague at Harvard, influential population geneticist Richard Lewontin, was very critical of what he perceived as attempts to give scientific credence to racist ideologies. Gould was a lifelong critic of the whole field of sociobiology, which had made a big splash in 1975 with the publication of *Sociobiology: The New Synthesis*, by Edward O. Wilson (also a colleague at Harvard, ironically invited to join the faculty a few years earlier by Lewontin). Gould's objections to sociobiology were in part well founded on scientific grounds, but were also the result of an unapologetic ideological commitment that seemed to have priority over the science and that led Gould to some unorthodox moves, such as reviewing (negatively, of course) Wilson's book in two different outlets.

Gould's best and most controversial writing on the lack of scientific support for racism was *The Mismeasure of Man*, a disturbing (if at times misleading) history of how biologists have frequently produced research "demonstrating" the intellectual inferiority of various groups, chiefly blacks and women. The main problem with Gould's stance—a perennial danger for the public intellectual—is that he overstated his case, thereby risking loss of credibility. Carl Sagan, as we have seen, exaggerated the environmental dangers of the smoke rising from the burning of oil fields in Kuwait, but at least he had the grace to admit his mistake and even to use it for pedagogical purposes. Gould simply seemed to reject out of hand even the remotest possibility that human cognitive traits may vary from one population to another because of variation in human genetic material—a claim that would be absolutely uncontroversial were we talking of any other trait in any other organism. He also seemed oblivious to two more problems intrinsic in his advocacy. First, that environmental determinism is as pernicious as the genetic kind he sought to defeat. If it turned out that human beings really are "blank slates" at birth (and the evidence is clear that they aren't), then this opens up the possibility of using behavioristic techniques to control them at will, resulting in the sort of nightmarish society that Aldous Huxley famously portrayed in *Brave New World*.

The second problem with Gould's take on racism and science is that it really shouldn't matter at all for our ethical discourse whether science discovers innate differences between, say, men and women (some of these innate differences are plain for everyone to see, as I'm sure you've noticed). This is because of what philosophers call the "naturalistic fallacy," the idea—surely familiar to Gould—that one cannot simply move from "what is" to "what ought to be" without a rigorous justification of such a move. In other words, facts about human biology or psychology have no bearing on what we should think of as the intrinsic worth of a human life. We are different from each other for a variety of reasons, both genetic (think of a basketball player) and cultural (think of someone who inherits a large fortune from his family), but all of us are (or should be) treated the same when it comes to the law. What this means is that while it is certainly praiseworthy to point out the scientific fallacies committed by racists, one does not need to overstate the case and completely deny even the possibility of genetic differences among human groups (an "is") to advocate equal treatment of all members of the human species (an "ought"). Indeed, to exaggerate one's case against biological differences opens up the possibility that one is eventually proven wrong, since, after all, statements of fact are subject to empirical verification. What then? Would Gould have to embrace racism (an ideological position) if someone showed him enough evidence of genetic bases of intergroup or intergender differences?

Gould was also active in another area of public intellectualism: the relationship between science and religion. Here he proposed a term, NOMA (for nonoverlapping magisteria, i.e., mutually distinct areas of influence), to label anew an old concept in theology and philosophy of religion. The basic idea is that science deals with facts, while religion is concerned with matters of value; therefore, the two are independent "areas of inquiry," and the perennial strife between science and religion is based on a misunderstanding by some people of the nature of both. If only it were that simple! First off, notice that Gould in this case does make use of the very same fact-value distinction that he seems to neglect in the case of biological research with potentially racist or sexist overtones. It is interesting that the same philosophical concept may or may not be deployed by Gould depending on whether it is convenient to his argument. Second, he certainly realized that it is precisely because most religions flagrantly violate NOMA as a matter of principle that we have a problem: pretty much all the world's religions do include as a fundamental component of their creed some "creation story," that is, an account of how the world (and human beings) came about. If one is a young earth creationist, one isn't violating NOMA by chance by rejecting evolutionary theory; that rejection is at the very core of the creationist's belief, which means that the conflict with science is inevitable. Indeed, Gould's NOMA works precisely where it is not necessary: people who understand religious texts to be taken metaphorically, and therefore not to be read as science textbooks, are not the ones that are fueling the science-religion cultural war to begin with.

There is a symmetrical reason why NOMA doesn't work: notwithstanding the above-mentioned distinction between facts and values (which is actually questioned to some extent by some modern philosophers like W. V. O. Quine and Philippa Foot, though that's a story for another time), religion as a human cultural and historical phenomenon can be, and has begun to be, investigated by science. As one can imagine, when scientists put religious belief and faith itself under the microscope, this doesn't help the notion of NOMA, and yet religion as a cultural experience is of course a legitimate object of research. This problem arises for Gould because he concedes too much to religion to begin with, by defining it as *the* human endeavor that deals with matters of value. He seems to neglect the obvious fact that philosophy does this as well, without the additional stuff about origin stories, supernaturalism, and all the rest. Moreover, Gould gives too little to science, since he doesn't explicitly consider the possibility that scientists not only can be interested in religion as a social process, but may also legitimately ask how, during the course of evolution, human beings developed a sense of morality to begin with. This is not the same as science dictating moral choices (that would in fact be a flagrant instance of the naturalistic fallacy), but it is perfectly rational to think that a sense of right and wrong arose over evolutionary time in response to the type of societal life that humans and other primates happen to live. Indeed, primatologist Frans de Waal has made just such case on empirical grounds, based on his studies of social interactions among bonobo chimpanzees, and his research violates NOMA for perfectly good reasons.[17] Gould's rather solomonic solution to the science-religion issue, therefore, doesn't stand up to scrutiny because it is based on an overly simplistic understanding of both science and religion. Nonetheless, it was a valiant attempt, and exactly the sort of thing that one expects a public intellectual to do. After all, just because one is an intellectual it certainly doesn't follow that one is also infallible.

We have so far examined the allegedly declining role of the public intellectual, as well as briefly discussed two controversial scientists who were public intellectuals and who paid a personal price in their careers for their choice of engaging with larger issues than those presented by their respective technical disciplines. But there is a whole other dimension to how science is understood by the general public, and this deals with the rise and rapid evolution of a new player in the public arena: the so-called think thanks.

Think-Tankery: From Intellectualism to Spindoctorism?

The twentieth century has seen the rise and rapid evolution of so-called think tanks, a trend somewhat parallel to the perceived decline in the quality of public intellectual discourse discussed so far. Think tanks are now pervasive worldwide: according to Diane Stone, as of the year 2000 there were four thousand think tanks active in nations across the globe,[18] and that number—very likely a gross underestimation—has certainly gone up since. Despite their number and prominent daily presence in the news, there aren't that many sociological studies of think tanks, and even fewer critical analyses of their role in shaping public opinion. Indeed, there is not much agreement on what, exactly, defines a think tank to begin with.

For the purposes of this discussion, I will use the term "think tank" to refer to a specific kind of organization, namely, a private group, usually but not always privately funded, producing arguments and data aimed at influencing specific sectors of public policy. This definition, therefore, does not apply to most advisory groups established by a given government, nor to university institutes or centers, nor to groups set up to resolve specific problems within the normal operation of a corporation. Although I am generally skeptical of think tanks as a concept, and very critical of the operation of specific ones, I do not mean to imply that all think tanks are useless or pernicious, nor that the general idea cannot in principle be pursued in an honest and constructive fashion. But I am struck by how little critical evaluation of the phenomenon there seems to be, both in the literature on think tanks and more importantly by the media and politicians who are the primary direct consumers of think tanks' output.

Donald Abelson gives a useful historical perspective on think tanks.[19] According to Abelson, the concept went through four relatively distinct phases since its inception about a century ago. The first generation of think tanks appeared in the early 1900s and was the product of the preoccupation of a small number of rich entrepreneurs concerned with the necessity of providing sound, rational advice to the government at a time of increasing complexity of both domestic and foreign policy problems. Thus people like Carnegie and Rockefeller provided large permanent endowments to these groups, which made them essentially independent from government support (and, therefore, influence) as well as freed them from the necessity of continuously raising private money (again emancipating them from possible leverage by their donors). (If you perceive a contradiction in the idea of anti-intellectual captains of industry donating money so that the government could count on the best intellectual input available, well, I am with you, but so the world appears to work.)

That is how groups like the Brookings Institution and the Russell Sage Foundation came about. They operated according to a model of a "university without students," attracting scholars from across the political spectrum, sharing the ideal (if not necessarily the practice) that reason reaches across ideologies. The results of these efforts were significant in shaping American society during most of the twentieth century, for example producing a national budget system as well as studies on the causes of warfare.

The second phase of think tank history began after World War II, when the government realized the importance of supporting scientific research because of its obvious relevance to all things military. The National Science Foundation (not a think tank) was established then, and so were think tanks like the RAND (Research and Analysis) Corporation. This is the first worrisome development in the evolution of think-tankery, since it obviously created a direct link between the funding source (in this case the government) and the recipient of research outcomes and policy advise (also the government), thus violating the intentions of the people who established the first think tanks earlier in the century.

Be that as it may, a few decades later we witness another change in concept, with the appearance in the 1970s and '80s of what Abelson calls "advocacy think thanks." These are groups like the progressive Institute for Policy Studies and the Center for American Progress, the libertarian Cato Institute, and the conservative Heritage Foundation. These think tanks tend to be significantly smaller than their predecessors, are dependent on continuous support by a large number of relatively small donors, and, more importantly, they often (though not always) blatantly blur the lines between research and advocacy. It is hard to read a "report" from some of these outlets and not think that their "conclusions" were actually the premises from which the whole exercise started, a definite departure from the model of a university—with or without students.

This is apparently an open secret, as the director of a major policy institute told Abelson: "[Think tanks] are tax-exempt cowboys defying the sheriff with their political manipulations. They don't want to stimulate public dialogue, they're out to impose their own monologue."[20] Or as Leila Hudson put it: "These institutions have substituted strategy for discipline, ideological litmus tests for peer review, tactics and technology for cultures and history, policy for research and pedagogy, and hypotheticals for empiricals."[21] Another problem intrinsic to the modus operandi of think tanks was best summarized by an anonymous official quoted in "On Mediators: Intellectuals and the Ideas Trade in the Knowledge Society," by Thomas Osborne: "One of the dilemmas of think tankery is that you can either say something sensible, practical and useful and have six civil servants and their dog read it, or you can say something spectacularly silly and have the media cover it."[22] None of these quotes sounds exactly like ringing endorsements of the idea of think tanks.

The last development in think tank evolution is what Abelson calls "vanity think tanks." These are even smaller and more ephemeral operations, often set up by individuals in pursuit of short-term political goals, like Ross Perot's United We Stand or Newt Gingrich's Progress and Freedom

Foundation. I will not examine these any further because they are only marginally relevant to science, if at all, although they have demonstrated their ability to affect the outcome of elections, and therefore—indirectly—to influence science funding and education.

What has this to do with science? As we shall see, it is think tanks like the American Enterprise Institute that have the gall to bribe scientists so that they speak critically of reports about global warming, and it is the anti-evolution Discovery Institute, a think tank out of Seattle, that is guided by a mandate founded on "a belief in God-given reason and the permanency of human nature," rather than on the serious examination of scientific theories such as evolution.

Alina Gildiner goes into some detail about how think-tankery concerning science dovetails into spindoctoring rather than rigorous analysis of the problem at hand.[23] Gildiner quotes Laura Jones, whose edited book on risk management has been published by a think tank, the Fraser Institute, as stating—without any data to back the claim up—that "zealous anti-risk activists have heightened our intolerance for small risks," such as the number of deaths resulting from wheels detaching from transport trucks that were poorly maintained (in Canada, government regulation lowered that number from 215 in 1997 to 86 in 2000, and of course the statistics do not include people who got injured, sometimes seriously, from collisions with the errant wheels). Another "exaggerated risk" discussed by Jones's authors is the connection between secondhand smoking and cancer. Indeed, the spindoctoring goes as far as taking a judge's decision to rescind the Environmental Protection Agency's rule that acknowledged secondhand smoke as a carcinogen—a legalistic decision based on procedural matters—as "evidence" that the claim is scientifically unfounded.

While commenting on another book, written by authors Robert Lichter (president of the think tank Center for Media and Public Affairs and paid consultant for Fox News) and Stanley Rothman (director of yet another think tank, the Center for the Study of Social and Political Change), Gildiner comments that one would look in vain for nuanced discussions about what should influence public policy. Rather, "what is to be found is a rhetorical, agenda-narrowing usage of scientific language and methods."[24] For example, Lichter and Rothman comment with an authoritative tone on the link between cancer and air and water pollutants, despite the fact that neither one is a scientist and that their sources were almost twenty years out of date. Science progresses; ideologies tend to linger unchanged (and often unquestioned).

One doesn't really need to read technical articles about think tanks to get a good idea of what many (again, not all) of them are about. Take a look at the Web site of, for example, the Cato Institute (cato.org) and examine their timeline of actions and publications; it speaks for itself. In 1992 they published *Sound and Fury: The Science and Politics of Global Warming*, in which they state that "there is neither theoretical nor empirical evidence for a catastrophic greenhouse effect and thus no case for what Vice President Al Gore calls a 'wrenching transformation' of the American economy." The following year, they produced *Eco-Scam: The False Prophets of Ecological Apocalypse* and *Apocalypse Not: Science, Economics, and Environmentalism*. Jump to 2000 and you will find *The Satanic Gases: Clearing the Air about Global Warming*. Is there a common thread here?

Or visit the even more obviously slanted site of the Competitive Enterprise Institute (cei.org), which the *Wall Street Journal* named (without a trace of irony) "the best environmental think tank in the country." One of their "scholars," Robert J. Smith, proudly proposed the rather questionable concept of "free-market environmentalism," and accordingly CEI was among the first organizations to criticize, back in 1990, the Clean Air Act amendments because they would "impose a new regulatory burden that would lead to higher energy prices" (perhaps, but would they make our air cleaner and our quality of life better?). There is more: CEI in 1992 "advised" the Food and Drug Administration to approve recombinant bovine somatotropin, which is a bioengineered growth hormone. Now surely such recommendation would be accompanied by the further suggestion of labeling the resulting products so that consumer choice—that ultimate driver of market forces—could be openly exercised? Think again: the CEI argued that mandatory labeling of dairy products is "inappropriate" because it violates the First Amendment (which includes the right to free speech—of the cows?).

In 1996 the CEI folks outdid themselves, launching their Communications Project "with the aim of showing how 'values-based' communications strategies can help claim the moral high-ground for our side, by making the case that capitalism is not only efficient, but also fair and moral."[25] If this isn't a frank admission that the CEI isn't at all in the business of research but squarely in that of advocacy, it's hard to imagine what is. Moreover, in 1999 CEI published a monograph arguing that the dumping (not their term) of trash in low-income rural communities in Virginia is actually good for those communities' economies. One wonders how many fellows of the Competitive Enterprise Institute live in such communities. As recently as 2001 CEI helped persuade the Bush administration (which surely did not need much convincing on this issue) not to regulate carbon dioxide as a pollutant, despite the fact that the gas is universally recognized by scientists as the major contributor to the greenhouse effect. In 2002, CEI won a lawsuit against the FDA in federal court, invalidating the 1998 Pediatric Rule requiring drug companies to test some drugs on children, instead of automatically approving them for use on children if they had passed tests on adults only. To CEI's evident chagrin, Congress wrote the Pediatric Rule into law.

And in this age of global economies, the actions of think tanks of this sort go well beyond national boundaries: in 2003 a CEI fellow sent a letter to the Philippine president, Gloria Macapagal Arroyo, advising her government to "disregard environmental alarmists and to continue allowing Filipino farmers to grow bioengineered corn." Arroyo followed the suggestion, with apparently no input from her electorate and with predictable financial gains by the bioengineering firms that support CEI's operations.

Back on national turf, in 2004 CEI exploited the usually high-quality television program *Bullshit!* with Penn and Teller (as I already mentioned, Penn Jillette is a fellow of the similarly politically oriented Cato Institute), normally devoted to debunking the paranormal. The target? Mandatory recycling programs. In the same year, CEI associates argued that living on a McDonald's diet is good for your cholesterol and that there really is no reason to panic if one finds high levels of lead in one's water supply (as it had happened recently in Washington DC).

In 2005 the Competitive Enterprise Institute pushed for opening the Arctic National Wildlife Refuge to oil and gas exploration, which would cause a major environmental disaster with likely little impact on the economics of oil supply, a discussion that once again took front stage during the 2008 American presidential elections. In the same year, CEI filed a challenge to the constitutionality of the 1998 multistate tobacco settlement, while the following year they patted the back of the EPA for allowing human volunteers to be used in studies on the effects of pesticides. Finally, again in a spectacular admission that the think tank is simply not interested in serious research and scholarship, the CEI defended the right of op-ed columnists to be paid by interested parties for the spin they give to their pieces: "An opinion piece—whether an individual op-ed or a column—exists to promote a point of view by argument. It does not seek to establish a fact, but to win people over to a particular viewpoint or opinion." Indeed, but the reader usually assumes—wrongly, as it turns out—that it is the author's opinion that one is reading, not that of a government agency or of a private corporation that surreptitiously paid for what superficially looks to the reader like an unbiased assessment. Caveat emptor!

While there is clear, and rather disturbing, evidence of an increasing shift from research to advocacy in the sociopolitical phenomenon of think-tankery, how much influence do think tanks really have, and how do they exercise it? Abelson and other researchers have repeatedly pointed out that it is difficult to measure such an elusive property as political and social influence. Attempts have been made to compare one think tank to another based on quantifiable parameters such as the number of press releases they put out that are picked up by major news outlets; the number of media appearances, especially on television, of the think tank's fellows; and even the number of cabinet-level positions filled by people associated with a given think tank. But of course this scratches only the surface, because political influence can go undetected when it manifests itself as indirectly shaping the views and policies of representatives and senators at both the state and federal levels. Obviously, much more research on and critical evaluation of the entire think-tankery phenomenon is needed; meanwhile, the public needs to be weary of the opinions proffered by allegedly unbiased "experts" whose affiliation to think tanks is barely acknowledged by media outlets.

We have analyzed jointly the alleged decline of public intellectualism and the rise of think tanks not because the two are necessarily directly related (though to some extent they may be), but because they both affect the teaching and understanding of science in the public arena, and they both are at the very least symptoms of the kind of transformation of modern society that has had commentators from Postman to Chomsky worried for a long time. Scientists and people genuinely interested in science should be worried too. We turn now to the exploration of how contentious the public discussions surrounding science and pseudoscience can be, examining two cases in which all the players we have considered so far—intellectuals, think tanks, and the media—play crucial and complex roles in shaping and sometimes warping public understanding of the debate.

CHAPTER SIX

Science and Politics

The Case of Global Warming

Everybody talks about the weather, but nobody does anything about it.—Mark Twain

We're in a giant car heading toward a brick wall and everyone's arguing over where they're going to sit.—David Suzuki, biologist

"It is at least 90% certain that human emissions of greenhouse gases rather than natural variations are warming the planet's surface." This is the conclusion, reported by the BBC on 2 February 2007, reached by the Intergovernmental Panel on Climate Change (IPCC) in a report that was then a conservative summary of our understanding of how much and why the world's climate is being altered. For a scientist, there is no contradiction between stating that something is known with about 90 percent accuracy and to say that it is pretty much certain. And yet, statements such as this one from the IPCC encapsulate much of the sense and nonsense that characterizes complex debates about the social impact of science, where facts, projections, and theories discussed by academics and other experts have to square with the understanding (or lack thereof) of the general public and the often very different priorities and timeframes of politicians. After all, the IPCC's projections reach through the year 2100 and beyond, but the typical time horizon of a politician spans two to five years, rarely more.

Now, if you were told that there was a 90 percent probability of your house being wiped out by a storm, or even a 66 percent chance of that happening (that was the level of confidence in the human origin of global warming expressed by a previous IPCC report, released in 2001), you would be concerned, *very* concerned. And yet, as we shall see, there are people who have written large volumes on climate change and are very much in a position to influence public opinion and policies who quibble on percentages and the exact range of estimates, helping to confirm Mark Twain's quip of what we do (or, rather, don't do) about the weather.

In this chapter I will take an unusual approach to discussing the issue of global warming. Several books and countless articles are available to the reader for in-depth or up-to-date treatment of the issue, and a single chapter in a broader book like this one will simply not do. Moreover, I am here interested not in the global warming controversy per se, but in what it may tell us about the complex intertwining of science and political ideology. What we will do instead, therefore, is to examine in some detail just two books written on the issue and use them as examples of how complex science is presented to and (mis)perceived by the public. These are two of the most important books written on the topic, although they are at the opposite ends of the spectrum, both in terms of style and of content. Surprisingly, we will discover that just because a scientist writes hundreds of pages with thousands of footnotes, it doesn't follow that his presentation of the problem is closer to reality than a simple slideshow put together by, of all people, a politician! We will find out that, perhaps not so surprisingly, when science has a potentially large and costly social impact, "controversies" are created in the media that do not really exist in the academy, and that personalities and sound bites—not to mention political agendas—count more than thoughtful reflection. We shall also explore a strange, but by all means not unique, situation in which politically and financially biased media outlets publicly defend as obvious truths positions that have little resemblance to what thoughtful scientists who work in the field actually endorse. Welcome to the climate change wars.

Before we can appreciate the sniping on both sides of the divide, however, we need to have a minimal understanding of what global warming is supposed to be about. The basic science is actually fairly simple, despite the fact that the details are inordinately complex, leading to a significant amount of legitimate uncertainty about the specific mix of causes and consequences of climate change. Global warming is probably caused by an assortment of natural and so-called anthropogenic causes. The first kind of cause is not really being debated, since everyone agrees that the earth has been through periods of both warm climate (Greenland—which today is largely covered by ice—was given that name because it was a lush land during the so-called Medieval Warm Period) and cold conditions ("Ice Age" isn't just the title of a Fox movie). There is also little doubt at this point—even amongst most skeptics of global warming—that anthropogenic (i.e., human-caused) effects are contributing to the problem. The question is rather the degree to which human activities are contributing to the overall increase in temperature and, even more controversially, what will be the long-term effects of our actions under a variety of possible scenarios, from continuing business as usual to drastically cutting industrial emissions of greenhouse gases.

But as far as how the phenomenon of warming works, the mechanism is rather simple, and it really does work very much like a greenhouse. If you have ever been in a greenhouse, you might have noticed that it is significantly warmer inside it than outside, even in the winter, and even in the absence of any artificial heating system. The reason for this is that during the day the sun warms up the soil and internal structures of the house; the heat thus generated—which would normally radiate into the atmosphere—is trapped by the glass, causing the interior of the structure to be significantly warmer than its surroundings.

Now, imagine the same exact phenomenon at the scale of the whole planet. Obviously, there is no big glasshouse in the sky, but we do have an atmosphere, and carbon dioxide (CO_2) is a major component of it. CO_2 has a tendency to behave like the glass in the greenhouse: it traps the heat that radiates back from the soil and the oceans, warming up the atmosphere. This system has been in dynamic equilibrium for the last several thousand years (a dynamic equilibrium is not an oxymoron: it means that the atmospheric chemistry and temperature keep ranging within certain limits, without going too far in one direction or the other—sort of like a thermostat in an apartment). What has been changing lately, to be precise since the Industrial Revolution, which started in Britain toward the latter part of the eighteenth century, is that human beings have began pumping so-called greenhouse gases (i.e., gases that contribute to the greenhouse effect) into the atmosphere. While emissions of CO_2 from industrial activities are the main culprit, methane and nitrous oxide also contribute significantly. These gases come from the operation of power stations, various industrial processes, the burning of fossil fuels for transportation (e.g., while driving or flying planes), agricultural activities, and a variety of other sources, but they all contribute to the problem in the same basic way that a glass screen triggers a warming effect inside a greenhouse.

Another important point to appreciate is that, as with many other things, the greenhouse effect is not inherently "good" or "evil." Indeed, without it, life on our planet could not exist, because Earth's basal temperature (i.e., the one that would hold without the natural greenhouse effect) would be much closer to that of the planet Mars—definitely on the cold side. On the other hand, too much greenhouse effect would turn the earth into something like Venus: an extremely hot and equally inhospitable hell. The problem, in other words, is not the greenhouse effect per se, but the possibility that too much of it is being triggered by what humans are doing on a planetary scale. And yet, some people are skeptical of what they consider to be the exaggerated claims of a misguided bunch of environmental alarmists. It is to this possibility that we turn first.

The So-Called Skeptical Environmentalist

To be skeptical means to harbor reasonable reservations about certain claims (the keyword here being "reasonable"). It means to want more evidence before making up one's mind. Most importantly, it means to keep an attitude of openness, to calibrate one's beliefs to the available evidence. It is interesting, then, that one of the most controversial books on the environment published in recent years is Bjørn Lomborg's *The Skeptical Environmentalist*, boldly subtitled *Measuring the Real State of the World*. It is a ponderous book, at 515 pages and a whopping 2,930 endnotes! But is it good science? Is the author a skeptic in the positive sense of the word, or a far too optimistic denier of an otherwise worrisome reality? We will go into a bit of detail analyzing one chapter of Lomborg's book, the one specifically concerned with global warming, because it represents a good example of how science can be used to oversimplify a complex topic and how hundreds of pages and thousands of notes do not necessarily make good scholarship.

When he wrote *The Skeptical Environmentalist*, Lomborg was an associate professor in the Department of Political Science at the University of Aarhus in Denmark. By his own account, the book originated from the notes for a series of lectures he gave there in 1997 and was published in Danish in 1998 (the first English edition dates to 2001). Why is this background relevant to the examination of the book? For a few interesting reasons. First, and most obviously, Lomborg is not a climate scientist. In the preface to the book he clearly says so: "I am not myself an expert as regards environmental problems." Why, then, attempt to write a scholarly book about the "true" state of the environment, one might wonder? Moreover, the extremely short (by academic publishing standards) period of time between the lecture notes and the finished book, together with the ponderousness of the book itself, should raise a true skeptic's eyebrow. I have some experience writing books, both technical and for popular audiences, and I know how much work goes into these projects, especially when one is an associate professor, still at the early stages of one's career and overwhelmed with many other commitments. Neither of these observations, of course, should substitute for a serious analysis of Lomborg's claims, but—to paraphrase Carl Sagan's immortal words—should definitely set our baloney detector on alert. Let us then proceed with the specifics and see what Lomborg's position is and how he presents it.

He starts out the discussion on global warming by stating that "this chapter accepts the reality of man-made global warming,"[1] but that in his opinion the claims about it—like the news of Mark Twain's demise—have been greatly exaggerated. Lomborg's main criticism is that the various computer models used by atmospheric scientists to make projections about future scenarios of the earth's climate are difficult to trust for three reasons: they attempt to model an extremely complex problem with oversimplified math; they are based on unrealistic assumptions about technological change in the near future; and they are laden with political value judgments.

I will leave a discussion of the last two points for later, because they reappear continuously throughout Lomborg's chapter. The point about the simplistic math underlying computer models of complex problems is true, but trivial: that is what all computer models are, a simplified representation of a complex situation. Think of computer models as maps of a territory. What makes a map good is most certainly *not* that it is very complex; indeed, if a map of, say, New York City, were so complex as to represent details of every building, you would never find your way around the city (not to mention that you wouldn't be able to fold the map and carry it with you!). Of course, if your map only shows the major intersections and gives you no information about, say, bus and subway routes, then it wouldn't be a good model either. The point is that simplification is a necessary feature of models (computerized and otherwise), and the real question is whether the simplification still yields a reasonably informative representation of the real thing. Neither I nor Lomborg can say, since we are not atmospheric scientists, and we have to provisionally accept what the experts in that field tell us is the best that can be done (we'll talk more about the concept of expertise toward the end of the book). Moreover, the report of the Intergovernmental Panel on Climate Change, repeatedly criticized by Lomborg for relying on such simplified models, had been peer reviewed by hundreds of experts, while Lomborg's own book—though published by the prestigious Cambridge University Press—has not been reviewed by a single natural scientist (it was published by Cambridge under their Social Science series).

There is another important point to understand about complex models. They are not meant so much to *predict* reality as to provide the users (in this case, scientists, policymakers, and the public at large) with a series of scenarios likely to emerge out of "if . . . and . . . then" situations. IF the assumptions of the model are approximately correct, AND the system being modeled (in this case, the earth's climate) keeps changing according to certain parameters, THEN we are likely to observe this or that outcome, with probability p. Models, therefore, are at bottom quantitative aids to our thought processes, enabling us to better grasp complex problems. Ironically, given Lomborg's criticism, a major goal of modelers is precisely to make their underlying assumptions as clear and transparent as possible so that people can pick the models apart and see what happens if one changes the assumptions this way or the other. The picking, however, has to be done fairly and reasonably.

Lomborg's strategy throughout seems to concede a few points to his opponents here and there, while at the same time attempting to undermine the overall credibility of previous studies, or at least to sow doubts (sometimes reasonable, sometimes not) in the minds of his readers. For example, he says that "there is no doubt that the temperature of the late 20th century is greater than many previous centuries, but this cannot be taken as a simple indication of overwhelming global warming as we are also coming out of a Little Ice Age. The claim that the temperature is higher now than at any time throughout the past 1,000 years seems less well substantiated."[2] He is technically correct, as we are in fact coming out of a so-called Little Ice Age, but his own graph of the data shows remarkable convergence of estimates from various studies showing not only a steady increase in temperatures over time, but a recent steep rise that seems compatible only with an anthropogenic explanation.[3]

While discussing several models predicting future emissions, changes in temperatures, and changes in sea levels, Lomborg complains that the UN Climate Panel, the IPCC, examined too many of them, making the discussion "numbingly complex."[4] This from someone who wrote a book on climate change that comes with a truly numbing 2,930 endnotes! He then states that "the important question is not *whether* the climate is affected by human CO_2, but *how much*."[5] Fair enough, but all models, which attempt to provide scenarios for what will happen with no change, with changes focused on economic considerations, and with changes focused on environmental considerations, agree that human-caused increases in temperature will be *at the very least* 2.0°C by 2100, with a corresponding increase of 31 cm in sea levels—which is more than sufficient cause for worry. Yet Lomborg goes on to say that "if the temperature increase we have been able to observe until now is not due solely to global warming, this also means that global warming is less important [than it would be if it were the only cause]."[6] This is true, but notice the sudden switch of focus, apt to confuse the reader: while past warming (the observed one) is in fact likely the result of a mix of anthropogenic and nonanthropogenic causes, the projected warming (the modeled one) is based on models that usually take into account only human-caused warming, so that their estimates will become *even worse* if there are additional, natural causes of warming. This is a little like saying that since not all deaths in New York City are caused by crime, one should not be worried by a projected rising tide of violent crime, because the presence of multiple causes makes crime itself less important. Would you vote for a politician that made that kind of reassuring argument?

Lomborg also attacks computer models because "computers are number-crunchers, not crystal balls."[7] To make his point, Lomborg discusses a 1996 IPCC study showing that models simulating only greenhouse gases predict (slightly) higher temperatures than models including both greenhouse gases and aerosols (like sulphates), because the latter have a cooling effect due to the fact that they reflect light. This sounds like a minor technical point, but Lomborg turns it into a serious attack on the reliability of climate science modeling: instead of praising atmospheric scientists for improving their models by taking into account better data, Lomborg focuses on the discrepancy to insinuate that one cannot trust models because their results change with changing assumptions. Of course, had the scientists in question *not* updated their models they would have been accused of not attempting to improve the realism of their figures. Damned if you do, damned if you don't.

A few pages later a similar criticism of models is prompted by the fact that a more recent IPCC report improved on earlier models by incorporating better parameters to represent clouds in the simulations. Since the new model is a change from previous predictions, how do we know that they won't change it again? This is a borderline dishonest play on a fundamental and widespread public misunderstanding of the process of science, where revision of previous results is not only normal, but clearly a positive aspect of a self-correcting enterprise, and yet is often seen by nonscientists as a sign of waffling, readily leading to an attitude of general mistrust of scientific expertise as a whole.

Lomborg at times makes it sound like he is discovering deep contradictions in other people's research, as if climate scientists are either incompetent or part of a worldwide conspiracy. For example, he discusses the possible amplification effect of global warming generated by a warming of the troposphere (the lowest part of the earth's atmosphere), which would trap more water vapor, thereby increasing the baseline effect of additional CO_2. He presents graphs, taken from published studies modeling global warming and therefore well known to the atmospheric scientists he criticizes, showing that the troposphere has been warming at a lower pace than the surface of the planet. That being the case, the amplification effect will be smaller than if the troposphere were warming at the same pace as the surface. But this is expected, since the troposphere includes much more volume of air than the surface-level film, and, again, this is well-known to people working in the field. The attitude here seems analogous to that of creationists who "discover" that evolutionary biologists have rejected an earlier explanation for, say, the evolution of feathers, while the source cited by the creationists is, in fact, a paper published by evolutionary biologists themselves. There is no worldwide conspiracy of either evolutionary biologists or atmospheric scientists, just the normal, self-correcting way science proceeds in its quest for a better understanding of the world.

What is the upshot of all of this nitpicking on Lomborg's part? We find out that "the IPCC simple model systematically overestimates the warming and consequently the much-quoted range of 1.4–5.8°C would be more likely 1.2–4.8°C. . . . Generally, this points toward a smaller but definitely not negligible effect of CO_2 on the climate."[8] In other words, despite doing his best, the environmental skeptic has to admit that the IPCC got it pretty much right.

Indeed, even success stories are criticized by Lomborg, as when he briefly describes the case of the ozone layer depletion and how the international community acted quickly to address the problem (the same example that former vice president Al Gore—whose take on the problem we'll discuss below—often presents as a case where the world actually got its act together in the face of an environmental threat). He then goes on to state that "it is worth pointing out that the implementation of the CFC [chlorofluorocarbons, the cause of the ozone depletion] restrictions was strictly profitable."[9] The not-so-subtle implication being that Lomborg sees intervention in the environment in a positive light if it is economically advantageous, an understandable position for an economist, perhaps, but not necessarily a good idea in general. Moreover, after discussing the predictions of increased numbers of skin cancers and of damage to fisheries and agriculture if the ozone problem weren't solved, Lomborg adds: "however, these are global figures over the next 63 years [this is an odd number to pick; your baloney detector should go to yellow alert], which because of both the long time and the number of people easily get very large. While they clearly show that global intervention was justified, it is also important to get a feel for the actual, *personal* consequences of the thinning of the ozone layer" [emphasis in the original].[10] This is like saying don't get scared by all those statistics on smoking and cancer, because they are amplified by big numbers and long time periods; in reality, at a personal level, you still have a decent chance to smoke and survive until old age, if you are lucky.

Of course, the nitpicking game can be played by both parties, and we can indulge in a little bit of it at the expense of Lomborg: his figure 146, for example, shows a well-known correlation between changes in global temperatures over the last 130 years and the cycles of activity of the solar sunspots. Lomborg uses this correlation to repeat his argument that at least part of the observed warming is the result of natural causes. Let us set aside for the moment the non-negligible detail that there is very little understanding of the causal links between sunspots and global temperatures (and without a well-established causal link, a correlation is just an interesting but potentially misleading statistic). Lomborg glosses over the fact that the otherwise very good match between the two curves he shows (plotting the evolution of sunspot cycles and earthly temperatures) completely breaks down during the last few decades, with temperatures continuing to increase regardless of the solar cycle. He attributes this to an "emerging greenhouse gas signal,"[11] that is, to human-caused global warming! Indeed, Lomborg shows a graph that clearly illustrates that greenhouse gases started to take off dramatically precisely in the period when solar cycles and temperature fluctuations began decoupling—but of course Lomborg doesn't draw his readers' attention to this annoying detail. As Thomas Henry Huxley ("Darwin's bulldog," as he was known) put it, in science "many a beautiful theory was killed by an ugly fact." And yet, the father of that theory is often oblivious to the death of his child.

Another interesting aspect of Lomborg's attitude is that he repeatedly stresses a point made by all technological optimists: things will get better with time because technology will find a solution to the problem, trust us. His example? At the beginning of the twentieth century icebergs were considered "a major climatic threat impeding travel between North America and Europe."[12] But all it took was for us to invent jet liners and, voilà, no more *Titanics*. Hardly the sort of argument that belongs in a scholarly work. Lomborg is also optimistic in another crucial area: the economy. He rejoices when the results of the same sort of modeling exercise that he so harshly criticized before end up producing results along lines that Lomborg favors. For example, some econometric models forecast an overall increase in wealth in the world by the year 2100, with a reduction of worldwide income inequality. This, according to Lomborg, will actually prepare even the developing world to pay for the damage incurred because of the effects of warming. At this point the reader can be excused if he is a bit confused: are models generally untrustworthy, or can we pick and chose when to trust them?

Lomborg makes the (reasonable) point that the best way to minimize the effects of global warming through the greenhouse effect is to develop and shift to renewable, nonfossil forms of energy. But then he immediately criticizes the fact that the scenarios proposed by the IPCC attempt to project what would happen both in the presence and absence of such technological (and economic) developments, because it is "patently implausible"[13] that all scenarios are equally likely. Of course it is not plausible that all scenarios are equally probable, but the whole point of producing scenarios is to allow the public and policymakers to explore, by way of computerized thought experiments, what *could* happen if we were to pursue one course of action or another, including less likely possibilities. The decision of what to actually do rests on all of us, not just on the modelers.

The really good news finally appears toward the end of Lomborg's book: even if global warming does happen, it's not going to be a big deal anyway. According to the author, by the time their agriculture will be hit by increased temperatures and CO_2 levels, "developing countries . . . will be considerably richer . . . and therefore more capable of handling the problems of the future."[14] Moreover, "new varieties of crops will be developed which are better able to exploit higher temperatures and higher CO_2 concentrations."[15] No need to worry too much about rising sea levels either, because "about three-fourths of this rise is due to the fact that the water has gotten warmer . . . and only one-fourth comes from changes in glaciers"[16] (and as we all know, it makes a huge difference for coastal communities to know why exactly they'll go underwater in the near future). Furthermore, "since the world also gets much richer, it can afford more protection" considering that "developing countries will be at least as well off as the industrialized countries are today."[17] At least. Moreover, "it seems likely that rich countries . . . will protect their citizens at such a low price that virtually no one will be exposed to annual sea flooding."[18] This was written only five years before hurricane Katrina devastated New Orleans and surroundings, and—needless to say—Lomborg offers absolutely no evidence for his claims about increased wealth and diminishing disparity between first and third world countries over the span of the next few decades.

Lomborg is not too worried about the projected consequences of global warming for human health, either, for example death from heat stroke. In by now typical fashion, he says that "a much richer world will be far more able to afford most people access to air conditioning."[19] I guess that was why in 2003—a mere two years after the publication of the book—thirty-five thousand people died in Europe (not in central Africa) during a heat wave,[20] while according to the *New York Times* (9 August 2006) 140 people died in California alone for the same reason three years later. And for an eternal optimist there's always the other side of the coin to look forward to: "in a warming world, there would also be fewer people dying from cold weather," except that Lomborg admits a few lines later that "winter-associated mortality is less strongly associated with temperature"[21] (and therefore warming things up would not make much difference in that department).

By this point in the book Lomborg's march of triumph is unstoppable. Despite his overall attempt at denying most of the reality of global warming, he manages to find much silver lining if one really is forced to admit a warming trend: "an argument could be made that overall such warming would be beneficial"[22] because we will have fewer cold-related diseases and better agricultural output. Moreover, while acknowledging that increased temperatures are translating into heavier and more frequent rainfall worldwide, which leads to more flooding, this also isn't too bad because "the combined increase in temperature, CO_2 and precipitation will make the Earth greener."[23] What are environmentalists complaining about? Thanks to global warming we can keep polluting *and* end up with a greener planet!

At times Lomborg seems to have a downright bipolar attitude toward global warming. While he keeps telling us not to worry, he also makes an argument that the 1997 Kyoto Protocol will do far too little to curb global CO_2 emissions (because it doesn't impose a cap on the emissions generated by developing countries) and therefore will have a small effect on rising temperatures. I must admit that his overall discussion of Kyoto got me to think hard for a bit, considering how the statistical projections and graphs he presents do in fact seem to indicate that even if the world adopted the protocol (most of it did, the United States being the major exception), it would be too little too late. Then again, I hadn't read the protocol myself, and apparently neither has Lomborg. When one eventually starts looking at the details, it turns out that Kyoto was not supposed to hold for the entire twenty-first century—as Lomborg erroneously maintains—but was meant simply as the starting point to get the notoriously recalcitrant community of world politicians to begin to take the problem seriously. Just as in the case of the ozone layer problem or of nuclear proliferation, one starts out with a timid first treaty with baby teeth and limited scope, and then moves up from there. It's the same recipe that gets most of us to tackle large tasks successfully: you begin with a small engagement and a good plan, and you ratchet up the effort on the wave of the first positive results. That's the way people successfully lose weight or get their finances in order (when they do). As they say, a long journey begins with a single (and often simple) step.

Throughout his book it is clear that Lomborg is mostly concerned about the economy of whatever environmental threat he writes about—again, understandable from the point of view of an economist, but not broad enough when we are talking about the welfare of the planet. It is not surprising that he ends his discussion of global warming by talking about costs but, again, the logic is both fascinating and somewhat misleading. Lomborg presents an estimated cost of $4,820 billion to the world if we do nothing and let global warming happen. That would seem to be a pretty scary incentive to action, until you compare it with his estimate for stabilizing the global CO_2 output ($8,553 billion) or—even worse—to what he says it would cost to limit the total warming to 1.5°C (a staggering $37,632 billion). But now the previously optimistic Lomborg turns out to have suddenly misplaced his Panglossian glasses, neglecting to factor *any* technological improvement or economic incentive into his calculations. Worse yet, by putting a price tag on the do-nothing scenario, he makes it sound like dramatically altering the planet's atmosphere and weather system is just another economic transaction, which we can afford if we have to (assuming that his calculations are even in the ballpark, something about which there is room for doubt). No mention of quality-of-life issues, lives lost or displaced, or ethical considerations about permanently altering the geochemical cycles and biosphere of an entire planet just because we can't stop ourselves from driving cars and operating air conditioners.

At some point Lomborg's argument becomes almost comical (in an unintended fashion, to be sure), for example when he notes that discussions about global warming tend to display "a fervor more fitting for preachers of opposing religions" and laments that "this is an indication that the discussion of global warming is not just a question of choosing the optimal economic path for humanity, but has much deeper, political roots as to what kind of future society we would like."[24] Indeed, he goes as far as to chastise environmentalists for being worried about more than just global warming. Lomborg quotes with disdain University of California, Berkeley's physicist John Holdren (later director of the White House Office of Science and Technology Policy under President Obama), who pointed out that the major problem we have for the future isn't a lack of energy, but how we use it: "clean-burning, non-polluting, hydrogen-using bulldozers still could knock down trees or build housing developments on farmland."[25] Apparently, this broader concern is bad, because it doesn't focus narrowly on global warming and its economic impact. No thinking globally in Lomborg's universe.

Lomborg ends his discussion of the entire issue with his equivalent of "it's the economy, stupid": "If we want to leave a planet with the most possibilities to our descendants, in both the developing and the developed world, it is imperative that we focus primarily on the economy . . . This puts the spotlight on economic growth, especially in the third world, while ensuring a global economy."[26] But it is precisely because the economy has been the only factor driving our planetary choices that we have global warming and general environmental degradation to begin with, so to suggest that the problem ought to be solved by economic means only is equivalent to telling an alcoholic that all he needs to solve his problem is the right kind of booze.

Lomborg is representative of many, especially in the United States, who seem not to understand the idea that "market forces" are not a panacea; they are efficient (at most) only at managing financial problems (and the 2009 worldwide economic collapse seriously questions even that mild assumption), not broader issues of quality of life or social justice. The argument is often made that renewable energies are more expensive than, say, coal or oil. But the disparity is there largely because one is comparing an emerging technology with a well-established one, and new technologies have always needed some kind of boost to establish themselves (oil is still heavily subsided in the United States, not to mention the car industry). More importantly, cost simply cannot be the only criterion, even if the environmentally friendlier solution were to always remain more expensive. If your kids were seriously sick, would you go to an incompetent doctor because he was cheaper than a competent one? It is astounding how this simple sort of consideration seems completely alien to the minds of so many (although fortunately far from all) economists. As the American writer Edward Abbey appropriately put it: "Growth for the sake of growth is the ideology of the cancer cell."

I was about to hand the manuscript of this book to my publisher with some satisfaction for a job reasonably well done when Lomborg sent me back to the library by publishing another book on global warming: *Cool It: The Skeptical Environmentalist's Guide to Global Warming*. It turns out that even the environmental skeptic is capable of evolving his thoughts about this issue, though Lomborg is not quite ready to join the growing consensus in the scientific community. In his new book, the Danish economist acknowledges much more anthropogenic effects than he was willing to concede before, and he grants that global warming is a serious problem that will take decades or even centuries to solve. But he still maintains that it isn't (too much of) a big deal.

Lomborg's new book relies on cost-benefit analysis, a fundamental tool of economists, as the judge of how much we should care about environmental, or any other, issues, and it puts forth three main ideas: yes, the planet is warming, but it is not going to be a catastrophe; it would cost too much to fix it by cutting greenhouse gas emissions; but governments should nonetheless invest significant amounts of money on new, cleaner energy technologies. This hardly sounds like the same person often quoted by Republican senator James Inhofe of Oklahoma, according to whom global warming is a myth invented by the liberal media and environmental fanatics. Yet Lomborg still insists in downplaying the danger and in rejecting what most economists see as the sensible thing to do: increase taxes on current energy use as a way to both pay for the damage being done to the environment and foster private investment in new technologies.

Cool It still features plenty of factual mistakes. For instance, Lomborg claims that very little or no melting has occurred on the surface of Antarctica. The data from a NASA study released in 2008 actually show newly formed (from melting) lakes the size of the state of California.[27] Despite this, I am inclined to go with Eban Goodstein, a professor of economics who reviewed the book for *Salon*: "Give Lomborg his whole argument. Suppose, as he believes, that Kyoto-level controls will cost a cumulative $5 trillion over the next hundred years. That is about two years' worth of increase in global output. Suppose also that we ignore Lomborg's advice and in the next few years freeze global warming pollution in the rich countries. That would mean that a century hence, our descendants, living in a much richer world, would have to wait an additional two years—until 2109—until a growing global economy left them as rich as they otherwise would have been in 2107."[28] In other words, what turns out to be "not a big deal" is the cost to curb greenhouse emissions, not the dangers of global warming.[29]

It is now time to turn to someone who thinks that global warming is in fact a very big deal and who, in perfect anti-Lomborg fashion, insists that we should do whatever it takes to reverse it.

The Crusading Politician

Al Gore, as he wryly says in a sequence of the movie *An Inconvenient Truth*, used to be the next president of the United States. Depending on your political leanings and worldview, that may or may not have been a good thing, but it would certainly have been a different sort of presidency from that of George W. Bush, particularly concerning the environment and the global warming controversy. At least, that is most certainly the impression one gets from the movie and the companion book of the same title, both based on Gore's itinerant multimedia lecture on the science and politics of global warming.

While *An Inconvenient Truth* presents a completely different story from *The Skeptical Environmentalist* and *Cool It*, in a sense it is just another problematic example of how environmental issues are discussed in the public arena. Gore's book is the antithesis of Lomborg's not just in viewpoint but also in attitude and content: where the Dane twists the science to fit his incurable optimism, the American sticks pretty much to the facts and to reasonable interpretations (with a couple of exceptions); however, Lomborg's thousands of notes and hundreds of references are countered by exactly none of either in Gore's book, making *An Inconvenient Truth* little more than a long, and largely uninformative, slideshow. To put it in another fashion: while Lomborg misinforms his readers, Gore gives them very little valuable information to go on, with the result that two of the most talked-about books on the environment of the last several years are most decidedly no starters when it comes to learning either facts or (reasoned) opinions about global warming.

Gore's book begins and ends with gorgeous satellite pictures of Earth, and it is in fact largely made of a collection of pretty pictures, some informative (a few data graphs here and there), some pertinent to the topic at hand (Earth, several hurricanes, again from satellites), and some from Gore's family album—something that has irritated his detractors and endeared him to his supporters, but that of course is quite irrelevant to the discussion on climate change. The book really does "read" like a slideshow, with entire page spreads occupied by large-font quotations, peppered by the occasional joke ("Our own Glacier National Park will soon need to be renamed 'the park formerly known as Glacier'"). Some of the images are stunning and effective and ought to be widely appreciated. For example, Gore shows his readers the border between Haiti and the Dominican Republic. The contrast couldn't be starker: the left side of the photo (Haiti) is a barren desert, while the right side (Dominican territory) is a luscious forest. The point: political decisions and the public policies they implement really do make a difference, despite widespread cynicism stating that all politicians are shortsighted and corrupt.

Equally stark is an entire series of "before and after" pictures of glaciers worldwide, where in each case it is simply astounding to see how fast the ice has been retreating over the last few years. The above-mentioned photos of hurricanes are also compelling, though that is one of the things in the book that gets Gore in trouble with experts in the field. As Eric Steig, a geochemist at the University of Washington, told the *Chicago Sun-Times*, atmospheric scientists do agree that global warming is likely having the general effect of producing more powerful storms (because warmer water increases wind speed).[30] However, one cannot point to individual events—like Hurricane Katrina, which devastated the Gulf Coast in 2006—and attribute them necessarily to the same global trend. As Steig put it, "Never in the movie [or the book] does he [Gore] say: 'This particular event is caused by global warming.' " No, but the implication of those photos within the context of Gore's discussion is clear (and misleading) enough.

On the other hand, Gore does deal quite effectively with some of the standard criticisms of global warming. For example, one of his graphs plots the temperatures from 1000 to 2000 CE, showing both the so-called Medieval Warm Period and (unlabeled in his book) the Little Ice Age that ended in the middle of the nineteenth century. It is clear that, contrary to contentions of global warming skeptics, those two fluctuations are mere blips on the screen compared to the post–Industrial Revolution rise in temperatures. Perhaps the most effective—and controversial—of Gore's graphs comes a few pages later: it is the record of both temperatures and CO_2 concentrations during the last six hundred thousand years.[31] The two graphs are clearly parallel to each other, and Gore in fact explicitly says that "they do fit together . . . when there is more CO_2 in the atmosphere, the temperature increases because more heat from the Sun is trapped inside."[32] The last part of the graph shows the current CO_2 level—clearly higher than anything over the last six hundred thousand years—and then projects the estimated CO_2 forty-five years from now: a steep ascent from 320 parts per million to above 600 ppm. Gore then, to great rhetorical effect, says: "about an inch in the graph [referring to the gap between temperatures during the last ice age and today] represents the difference, in Chicago, between a nice day and a mile of ice over your head. Imagine what three times that much on the warm side [the projected difference in forty-five years] would mean."[33] Indeed, A. O. Scott, in the *New York Times*, wrote of that graph, also presented in the movie version of Gore's lecture: "I can't think of another movie in which the display of a graph elicited gasps of horror . . . the effect is jolting and chilling."[34] Yet, Steig (again in the *Chicago Sun-Times*)—who has seen the Gore movie and thinks the vice president got the science largely right—pointed out that "Gore is correct to link temperature and CO_2 in ice core records. That's very sound science. But he is incorrect to imply that you can take the one curve and use it to predict where the other curve will go in the future. It ain't so simple."[35] In fact, Gore himself admits that "it's a complicated relationship," but the reader is led to one unmistakable conclusion nonetheless. How much is one allowed to stretch the truth (a word that prominently appears in the title of both the book and the movie) in order to make a rhetorical point?

While graphs or numbers are often used accurately and to great effect throughout the book, in other cases one wonders whether Gore remembers how to properly present information from his introductory college courses. For example, at one point we see three lines that are supposed to represent the concurrent increase of water temperature, wind velocity, and, therefore, moisture content of storms. This is all nice and dandy, except that the graph has absolutely no labels or scale on either axis; it's just three colorful lines going up in concert! What is an intelligent reader supposed to make of that? Then again, an excellent point is driven home when Gore tells us that the number of peer-reviewed papers dealing with climate change over the ten years previous to the book publication is 928; of these, exactly zero doubt an anthropogenic cause for global warming. During the same period, 636 articles on global warming appeared in the popular press, of which 53 percent doubt that global warming is a human-caused problem. While this could have been said in a couple of sentences (like I just did), Gore takes two full pages (again, slideshow style). On the other hand, the point is not to be missed, and in fact it applies generally to other "controversies," such as the one about the teaching of evolution, where again there is essentially no professional biologist who seriously doubts the theory of evolution, and yet the polls indicate about a 50–50 split in public opinion.

Perhaps two of the most useful contributions of Gore's book are found in the last few pages. I have summarized them below in table form for ease of reference by the reader. The first one is a compendium of the most common myths about global warming that are widespread among the general public and politicians alike:

Myth	Why Is It a Myth?
Scientists disagree on whether global warming is caused by human activities.	One can always find individual scientists who disagree on anything. But there is an overwhelming consensus among climate scientists (those who actually understand the problem) that human activities are, in fact, a major reason why global temperatures are rising.
Since many factors affect global climate, it is not fair to focus on CO_2.	Yes, multiple factors affect every complex system, such as global climate. However, the fact remains that increasing levels of CO_2 are a major determinant of warming. To discount this would be like saying that since crime is caused by several factors, it doesn't help to, say, improve education in poor neighborhoods (a major factor known to be linked to crime rates).

Myth	Why Is It a Myth?
Climate varies according to natural cycles, so this isn't our fault.	This is a perfect example of logical fallacy. It is true that there are natural cycles of variation in global climate, but it doesn't follow that the current trend is therefore not caused by humans. The data clearly show that the level of CO_2 in the atmosphere has dramatically increased after the beginning of the Industrial Revolution, and the degree and speed of increase cannot be explained solely by natural variation.
The hole in the ozone is what really causes global warming.	While the ozone hole is caused by human activities (in particular the release in the atmosphere of chlorofluorocarbon chemicals), it has nothing to do whatsoever with global warming. The two phenomena are not causally related at all.
It's already too late.	No, it isn't. Even global changes such as those that we are inflicting on the planet are reversible. Of course, the more we wait, the more difficult they will be to reverse. To use an analogy from everyday life, if you want to lose weight, you can start any day, but the more pounds you put on, the more difficult it will be and the more drastic (and painful) measures you will have to take.
Antarctica's ice is growing, so there is no global warming.	Some areas of Antarctica's ice sheets are in fact growing, but the overall trend, both within that continent and worldwide, is of a clear decrease of ice sheets. This objection confuses local with global trends—the latter are the relevant ones as far as the greenhouse effect debate is concerned.
Global warming is good, winters will be less cold and plants will grow better.	Again, this is an example of confusing local with global effects. Some regions of the world will undoubtedly benefit from (some) global warming, at least temporarily, but the overall impact on a planetary scale will be negative.
The warming effect is really due to cities trapping heat.	This is a simple case of misinformation. Scientists are aware of the so-called urban heat island effect, and they measure temperatures of cities in parks (which are cooler), not on the asphalt. More importantly, the warming trend affects both rural and urban areas, not just the latter.
Global warming is caused by a meteor that crashed in Siberia in 1908.	This is so silly it hardly deserves a response. First, that impact was much too small to cause a global effect. More important, large meteors affect climate only temporarily, and there is no way a meteoritic impact could keep increasing temperatures a century after the crash.
Temperatures are not going up everywhere, so global warming is a myth.	Yet another example of confusion between local and global effects. The climate is a complex system, so an increase in greenhouse gases (not only CO_2) has multiple and diverse local effects, depending on geography and other considerations. But the global trend is undeniable.

The other list is perhaps even more important. Gore realizes that an honest book (or movie) on global warming cannot but leave the sensitive reader (or viewer) both overwhelmed and depressed by the extent of the problem. Therefore, it is useful to know what one can do, immediately, to help out:

Where and When	What You Can Do, in Practice
In your home	Choose energy-efficient lighting. Choose energy-efficient appliances. Properly operate and maintain appliances. Heat and cool your house efficiently, paying particular attention to your insulation. Conserve hot water. Turn off appliances that you don't use. Switch to green power if available from your utilities company, or ask them to provide it.
When you travel	Drive less, use a bike or public transportation instead. Even better, walk! If you have to drive, develop a style that saves fuel (it's also good for your and your family's safety). If you have to buy a car, go for a fuel-efficient model or, better yet, for a hybrid. Use alternative fuels (like biodiesel), if your car can process them (and chose a car that can, if possible). If you can, telecommute, that way you won't need transportation to work at all! Reduce air travel to a minimum, and if you do have to travel, purchase "offsets" (made available by several airlines and travel Web sites) so that your "carbon footprint" is reduced.
When you make everyday choices	Consume less of whatever you do consume. Buy things that have more durability. When you buy, especially groceries, bring your own bags— don't use plastic from the vendor. Recycle whatever and whenever you can. Reuse paper, and use as little as possible of it to begin with (after all, we are in the electronic era of the paperless office!). Buy things that are wrapped in as little packaging as possible. Ask your favorite merchants to prefer merchandise that is not overwrapped. Compost! Don't buy new bottles of water; bring a refillable one with you. Eat less meat (this isn't an appeal to vegetarianism, it's simply that producing meat for human consumption is much more environmentally damaging than producing vegetables and fruits). Do not eat endangered species (like Chilean sea bass). Do some research on environmentally friendly Web sites, some of which make handy downloadable restaurant and groceries guides available. Buy locally produced goods; not only will that help your local economy, but it will reduce global transportation of goods, a major source of environmental pollution.
Education time!	Read and learn more about environmental problems. (Though, unfortunately, not from either Lomborg's or Gore's books.) Talk to others about what you learn. Encourage your workplace and your children's school to recycle and reduce waste.

Where and When	What You Can Do, in Practice
	Vote with your dollars by buying products from environmentally conscious companies.
	Investigate where your investment money goes—you may not like it. There are specialized portfolios for people who wish to invest in environmentally conscious companies.
	Take political action and, of course, vote!
	Directly support an environmental group. Money does make a difference.

I cannot neglect ending this section by mentioning that Gore also wrote a book since *An Inconvenient Truth*, though his target now is the general degradation of reasoned discourse in the United States, particularly in American politics. *The Assault on Reason* is a passionate defense of the role of rationality in the public arena, just about the same motivation that led me to write the present volume. Gore's new book is not about global warming per se, but addresses what concerns us next: how are we to make up our mind in an informed fashion when it is so easy to pick and read only those opinions one already agrees with?

The Public Response: How Do You Tell Who's Right?

One way to answer this question is to examine the record about how Lomborg's and Gore's books have been received by commentators and experts. Things quickly get both fascinating and a bit disquieting. The pattern emerges clearly and distinctly from even a cursory analysis of the available reviews: *The Skeptical Environmentalist* has been hailed almost universally by conservative newspapers and financial publications, while it has been severely criticized (along similar lines to what I have outlined above for the specifics of Lomborg's chapter on global warming) by prestigious science magazines such as *Nature*, *Science,* and *Scientific American.* Gore's book (and, even more, the movie), on the other hand, has encountered an almost symmetrical fate: praise by experts, vitriolic criticisms from conservative quarters. Consider some examples.

Let's begin with *The Skeptical Environmentalist*. Matt Ridley (who also wrote about it for the *Economist*) said in the *Sunday Telegraph* that it is "probably the most important book on the environment ever written," lambasting the environmental movement because "the alarmists were badly wrong . . . The extreme greens have been so wrong for so long that you would think somebody might have noticed."[36] Andrew Greely, in the *Chicago Sun-Times*, wrote "I will not dialogue on the environment with anyone who has not [read it]," and goes on to charge environmentalists of having "helped create a new religion whose devotees are compelled to accept false prophecy as unquestionable truth"[37] (how exactly this compulsion would occur he neglects to specify). There is more. In the *National Review*, Jonathan Adler says that the book "represents the most substantial challenge to the green orthodoxy that modern civilization is producing environmental ruin"[38] (notice, again, the religious analogy). Finally, Charles Rubin, in the *Weekly Standard*, rhetorically asks, "Why don't the facts seem to count for much in environmental matters?"[39] Fortunately, Rubin says, Lomborg comes to the rescue: "[he] skillfully uncovers the source of the bias towards bad environmental news in the scientific community, environmental interest groups, and the media." So "the media" are in on the great environmentalist conspiracy, never mind that all the above-cited publications are an important component of "the media."

After digesting these reviews, the intelligent reader may be excused for coming to the conclusion that Lomborg has really nailed down the problem and that his uncompromising scientific analysis of the facts has dissipated the fearful phantoms conjured up by environmentalist nightmares. What, then, do real scientists think of the book? *Scientific American* actually devoted a special section of an issue of its magazine to a series of in-depth critiques of *The Skeptical Environmentalist*. Here is a sample from an article by Stephen Schneider: "[in his] preface, Lomborg admits, 'I am not myself an expert as regards environmental problems'—truer words are not found in the rest of the book."[40] Michael Grubb, contrary to Lomborg, is in fact a specialist on climate change. In the prestigious *Science* magazine, he writes, "I can only describe his [Lomborg's] analysis [of climate change] at best as inconsequential . . . He shows no appreciation for the practical or moral dimensions of impacts on potentially billions of people."[41] The other world leader in science publishing, *Nature*, put out a review by my former colleague at the University of Tennessee Stuart Pimm, written in collaboration with Jeff Harvey. According to Pimm and Harvey, *The Skeptical Environmentalist* is "a mass of poorly digested material, deeply flawed in its selection of examples and analysis." While discussing his litany of environmentalist complaints, continues the review in *Nature*, "[Lomborg] quotes news magazines and a book by two science-fiction writers, but not scientists directly . . . Out of around 2,000 references, about 5% come from news sources and 30% from web downloads,"[42] which is not a good record for what is supposed to be a scholarly publication.

The situation isn't much less confusing when we move to examine the reception of Gore's movie and book. Let's see if you can spot the conservative-leaning authors and organizations in what follows. Simon Grose, writing in the *Canberra Times*, begins his article by acknowledging—grudgingly, one imagines from his tone—that "Al Gore must be a better salesman than me."[43] The *New York Times* reported that the Competitive Enterprise Institute (a libertarian think tank) produced two television commercials defending CO_2 (the first time, I think, someone paid good money to defend a gas!) with the hilarious (if not taken seriously) line "Carbon dioxide. They call it pollution. We call it life."[44] The same video, according to the *Chicago Sun-Times*, features Gore riding a cartoon airplane to the tune of the Allman Brothers' "Ramblin' Man," explicitly accusing the former vice president of hypocrisy for telling us to reduce our carbon footprint while he merrily goes flying around the planet to trumpet his extremist environmentalist message.[45] What the good people at the Competitive Enterprise Institute neglect to tell viewers is that Gore pays for offsets to his travels in order to achieve a zero-carbon balance, just as he encourages the readers of *An Inconvenient Truth* to do.

Conservative columnist Charles Krauthammer, quoted in the *New Yorker* (itself certainly not a conservative outlet), quipped that "it looks as if Al Gore has gone off his lithium again,"[46] referring to a typical medication for people with bipolar disorder. And Fox-TV commentator Sterling Burnett—according to the *Chicago Sun-Times* article cited above—made a direct comparison between learning about the environment by watching Gore's movie and learning about Nazi Germany from one of Goebbel's propaganda films during World War II (it is by now a well-established fact in rhetorical discourse that to compare one's opponent to Hitler and his minions is the last resort of the rhetorically inept). It is worth noting, incidentally, that the Competitive Enterprise Institute is funded in part by Exxon-Mobil, the oil giant, which also sponsors Fox-TV and pays for some of its staff. Coincidence or intelligent (albeit devious) design?

On the flip side, scientists have been fairly positive about Gore's message, with the *San Francisco Chronicle* reporting that nineteen climate scientists who have read the book or seen the movie told the Associated Press that the science in them was accurate.[47] We have examined above some specific criticisms moved by experts to individual statements in the book or movie, but those are to be expected for what is not, after all, a scholarly work written by a scientist.

What is one to make of these stark contrasts? This is much more puzzling than the typical situation in which different opinions are being voiced in the media on a particular subject matter. First of all, the issues being debated are—at face value—factual, and therefore amenable, at least in principle, to being settled one way or the other. Second, it is obvious that the relevant divide is not merely between left- and right-leaning press, but between part of the press (the one with financial and social incentives to be favorable to Lomborg and critical of Gore) on one side and the vast majority of scientists on the other. Again, the similarity with the so-called creation-evolution controversy is striking. There too, the controversy is such only in the public's eye, fueled by the media, while scientists all agree that there is nothing particularly troublesome in the theory of evolution.

Under these circumstances, it is difficult to make up one's mind about what is really going on. Even a technically savvy, intelligent reader may have trouble assessing the competing claims, and few people have the time or stamina to read Lomborg's book or the complete report by the Intergovernmental Panel on Climate Change. Nonetheless, as citizens of the world, and in particular of democratic nations, we will be called to vote—both directly at political elections and indirectly by a variety of other means—on the issue of climate change. How shall we do it?

My particular opinions should be clear by now, but I will state them unambiguously just in case. I think humans are in fact contributing to a significant alteration of the world's climate, even though we are certainly not the only factor at play. I am convinced that the long-term consequences will not be pleasant and may in fact be disastrous. I recognize that even the best science is only capable of giving us provisional answers and that the best forecasts for the future carry a good degree of uncertainty. But I also think it would be foolish for society not to be seriously proactive in this area, considering that, after all, we have only one planet to live on.

Of course, the reader shouldn't take my word for it. Yet, while attempting to make up one's mind on this complex topic, one should consider what the scientists say (they overwhelmingly agree that there is an anthropogenic effect on the world's climate and that we ought to do something about it) and what international agencies such as the IPCC propose and why (with the caveat that they tend to be more moderate because they have to balance the science against the political and economic interests of the member states). It is also important to consider opinions published in media outlets, of course, since those outlets are supposed to be the mediators between the rather rarefied world of experts and the rest of us. However, it would be naïve not to take into account the fact that the media come in a variety of flavors, some more or less independent, others clearly subscribing to a particular political-ideological framework, and in many cases—most obviously for think tanks such as the above-mentioned Competitive Enterprise Institute—reflecting both specific ideologies and precise financial interests.

So, what's an interested, politically moderate but socially concerned citizen to make of all of this? Who is close to being right and who is way off the mark? Perhaps the best one can do at this point is to take as a benchmark the 2007 report of the IPCC, with the proviso that it is more moderate than most scientists would prefer (and much more forceful than conservative think tanks and media outlets would like). This report was even endorsed by the administration of George W. Bush, who for years has denied that there is a global warming problem and has refused to subscribe to the Kyoto Protocol.

The IPCC's report, despite its moderation, is no comfort to Lomborg-style skeptics of global warming. It concludes that "global atmospheric concentrations of carbon dioxide, methane and nitrous oxide have increased markedly as a result of human activities since 1750 and now far exceed pre-industrial values determined from ice cores spanning many thousand of years."[48] The IPCC states that warming of the climate system is "unequivocal" and that its human origin can be stated with "very high confidence."[49] So much for the likely causes of the observed trends (i.e., the facts). As for future scenarios, the IPCC is only partially optimistic: "Even if the concentrations of all greenhouse gases and aerosols had been kept constant at year 2000 levels, a further warming of about 0.1°C per decade would be expected"[50] because the climate is a complex system in which, once major changes are set in motion, effects can keep unfolding for some time after the original causes have been removed. And, needless to say, the world has most certainly *not* frozen emissions at 2000 levels.

Moreover, "continued greenhouse emissions at or above current rates would cause further warming and induce many changes in the global climate system during the 21st century that would *very likely* [italics in the original] be larger than those observed during the 20th century."[51] As for long-term changes that could affect humanity, "anthropogenic warming and sea level rise would continue for centuries due to the timescales associated with climate processes and feedbacks, even if greenhouse gas concentrations were to be stabilized."[52] And these, remember, are the moderate scientists! The IPCC's report is not a counsel for despair, but a difficult attempt to tell the rest of us what is likely to happen depending on a variety of possible scenarios, over which we do have at least partial control.

And now for a final twist that I couldn't have better written into the story if this were a work of fiction rather than a discussion of science and politics. The day the IPCC report was released (2 February 2007), the British newspaper the *Guardian* ran a story alleging that another think tank, the conservative American Enterprise Institute, had actually attempted to bribe several scientists and economists (to the tune of $10,000 plus travel expenses) to write negatively about the IPCC report.[53] The AEI receives money from—you guessed it!—the Exxon-Mobil oil giant and was at the time deeply entwined with the Bush White House. Do I need to say more to push your baloney detector all the way up to red alert?

CHAPTER SEVEN

Science in the Courtroom

The Case against Intelligent Design

A Darwinist cannot invoke angels adding staples to [mouse] traps, because the angels are on *our* side.—Michael Behe, biochemist

I cannot persuade myself that a beneficent and omnipotent God would have designedly created parasitic wasps with the express intention of their feeding within the living bodies of caterpillars.—Charles Darwin

The evolution-creation controversy is one of those perennial issues that never seem to go away no matter how much scientists alternately ignore or confront them. Even more so than in the case of global warming that we just examined, there is no real controversy about this among scientists, who overwhelmingly accept the theory of evolution as by far the best available explanation of life's history and diversity. Indeed, unlike the case of global warming—which was, in fact, legitimately debatable until recently—such consensus about evolution developed almost a century ago, and has hardly changed since. Yet, similarly to what Al Gore pointed out for climate change, the American public is split down the middle with regard to evolution—a sharp contrast between what scientists think and what the laypeople perceive.

The evolution cultural wars make for a splendid case study in the quest for what distinguishes science (evolution) from pseudoscience (creationism). But this chapter will tackle the issue from a different perspective, examining in depth the outcome of a landmark trial in a court of law in which a judge was asked to decide the very question that is at the center of this book: what is science, and what is pseudoscience?[1] Judge Jones in Dover, Pennsylvania, delivered a detailed opinion that includes a thorough examination of the history of the controversy, the science behind both evolution and intelligent design theory (ID, a form of creationism), and even the philosophy of the divide between science and nonscience.

But before delving into the fray, we need as usual to arm ourselves with a preliminary understanding of what evolutionary theory itself is all about. As in the case of the greenhouse effect considered earlier, the basic idea is not that difficult, though the details can be extremely complicated, especially in the mathematical version of the theory. The modern theory of evolution began with Charles Darwin and his 1859 book, *On the Origin of Species by Means of Natural Selection*. Darwin's book was, as he himself put it, "one long argument" to make essentially two points: first, all life on Earth is related by common descent; second, the major explanation of the diversity of life, and especially of the obvious adaptation of living organisms to their environment, is a process that he called natural selection.

The evidence that Darwin amassed in favor of his theory included an examination of the fossil record, with its traceable change in forms from simple to complex over long periods of time; the comparative anatomy of organisms such as vertebrates, clearly showing that they share similar structures (e.g., the forelimbs) that can be adapted to different functions (e.g., walking vs. flying); the study of animal breeding, demonstrating the rapid change in form that can be obtained even by unconscious selection (although breeders select directly, and therefore consciously, for certain characteristics, other traits change as an unconscious byproduct of this selection); the biogeography of both living and fossil forms, which can be used to trace the change of related species over time, as well as their adaptation to changing environments.

Modern biology has greatly expanded the same lines of evidence and added a wealth of information that was not available to Darwin from developmental biology (species that are considered to be closely related also share similar developmental systems) and molecular biology (the similarity in DNA sequences of different species matches the one predicted on the basis of anatomy and biogeography). To put it simply, the empirical evidence supporting Darwin's fundamental insight that living organisms changed over the course of Earth's history is as solid as the evidence in any open-and-shut murder case, or, more pertinently, as solid as any other currently accepted mainstream scientific theory, such as Einstein's general relativity.

While many empirical and theoretical advances have been made in biology after Darwin, and while biologists today recognize a variety of evolutionary mechanisms besides natural selection, it is crucial to understand two ideas that are so often obfuscated within the context of the evolution-creation debate. First, common descent is a pattern (not a mechanism), which is explained by a variety of causes (mechanisms), the major one being that living organisms possess genetic material that they pass from one generation to another. Second, and most crucially, natural selection is not a random process. The latter misconception is so widespread, even among educated people, that it is worth restating the point unambiguously: natural selection is an evolutionary mechanism that is not random, and therefore one cannot reasonably say things like "evolutionists think that humans came about by chance." Nothing could be further from the truth.

Evolution by natural selection, at its core, works like this: living organisms are characterized by heritable variation for traits that affect their survival and reproductive abilities. This heritable variation originates from the (truly random) process of mutation at the level of DNA (although Darwin didn't know this, he observed that there was variation and noted that somehow it was heritable from one generation to the next). Natural selection then *is* the differential survival and reproduction of organisms that have (heritable) traits that allow them to do better in the struggle for life. That's it, but the concept is profound because the process of evolution turns out to be largely the result of two components: mutations (which are random) and natural selection (which, again, is not random). It is the joint outcome of these two processes that—according to evolutionary theory—explains not only the diversity of all organisms on Earth, but most crucially the fact that they are so well adapted to their environment: those that weren't did not survive the process. Indeed, many organisms do not survive or reproduce, with the result that more than 99 percent of the species that ever existed have by now gone extinct.

You may find it intuitively difficult to believe that two relatively simple natural processes can produce the complex order we observe in living organisms. But the beauty of science is that it so often shows our intuitions to be wrong. Moreover, there are other examples of order in nature arising from a mixture of random and deterministic forces, hurricanes being one. Even more simply, just think of the orderly distribution of pebbles on a rocky beach, with the small pebbles further up the shore and the largest ones distributed near the water. What accounts for that? Did an intelligent designer come by and manually sort all the pebbles? No, the pattern was produced by the combination of two causes: the (random) action of waves, which pushes all pebbles toward the shore, and the (nonrandom, selective) effect of gravity, which makes it so that the lighter pebbles on average end up further away from the water.

One additional important concept follows from our simple discussion so far: evolution is *both* a theory and a fact, contrary to simplistic creationist views. How can this be? Evolution is a fact in the sense that it is beyond reasonable doubt that living organisms have changed over time throughout the history of the earth. It is a theory in the sense that biologists have proposed a variety of mechanisms (including, but not limited to, mutation and natural selection) to *explain* the fact of evolution. Similarly, in modern physics, gravity is both a fact (apples really do fall on people's heads when they become detached from their tree) and a theory (according to Einstein, gravity *is* a local deformation of space-time caused by physical objects with mass).

We now have the basic tools to follow the bizarre story that unfolded in Dover, Pennsylvania, and culminated in one of the best examples of how science and philosophy of science can play a surprising and fundamental role in our courtrooms, helping to decide what should and should not be taught in our public schools by way of an analysis of the distinction between science and pseudoscience.[2]

The Simple Statement That Led to a Storm

On 18 October 2004, the Dover School Board passed the following resolution: "Students will be made aware of gaps/problems in Darwin's theory and of other theories of evolution, including, but not limited to, intelligent design. Note: Origins of Life is not taught." That decision came after almost three years of intense political maneuvering on the part of several members of the board, and it eventually led to a historical trial in which proponents of intelligent design were handed a devastating defeat by a conservative judge appointed by President George W. Bush. The 139-page decision by Judge John E. Jones III is worth reading in its entirety, and I will discuss it in some detail because it will guide us through a fascinating tour of human deception worthy of a mystery novel, all the while teaching us something about the nature of science and the difference between science and pseudoscience. It truly is a case study destined to become a classic in the cultural wars.

The story, as Judge Jones tells it,[3] began in January 2002, when Alan Bonsell, president of the Dover School Board, publicly declared that his two main goals for board action were to push the teaching of creationism in the district's schools and to reinstate public prayer. Both goals, of course, violate the constitutional separation of church and state and should have therefore never been on the agenda at Dover, but it is the ignorance and bigotry of local officials that often causes trouble where there should have been none.

The legal side of things is rather simple. The so-called establishment clause of the First Amendment to the U.S. Constitution (passed in 1791) reads: "Congress shall make no law respecting an establishment of religion, or prohibiting the free exercise thereof." The Fourteenth Amendment, passed in 1868 (i.e., after the Civil War), applies federal law to the states. There is very little room for interpretation of the establishment clause: it is made of two subclauses, the first of which prohibits the government from favoring or forcing religion—any religion—on the citizenry; the second prohibits the state from impeding in any way on the free (but private) exercise of religion. It is hard to imagine how anyone could therefore seriously argue that teaching creationism (which the Supreme Court had determined in 1987 is not science, but religion) or officially sponsoring prayer in schools (as opposed to students privately praying while on recess at school, which is of course perfectly legal) would not violate the First Amendment. And yet, failing to understand this represented the beginning of trouble for the Dover School District.

Bonsell eventually confronted teachers in the district in person about the teaching of evolution in the fall of 2003, an unprecedented administrative step that sent a chilling message to the teachers: stay away from the controversy if you don't want to trigger the ire of the board. This had an immediate effect on one of the teachers who testified at the Dover trial, Robert Linker.[4] Before the meeting with Bonsell, Linker used to tell his students that creationism is based on "religion and Biblical writings," which made it illegal as a subject matter in public schools. After the meeting with Bonsell, Linker dropped any mention of the controversy to his students and even stopped using helpful teaching material to aid students in making the distinction between science and religion. It was the first round of an escalating confrontation that would eventually completely vindicate the teachers and cast a serious cloud of misconduct on the administrators.

Enter another shadowy character: William Buckingham, whom Bonsell had appointed chair of the board's Curriculum Committee. In early 2004, Buckingham contacted the Discovery Institute, a Seattle-based "think tank" devoted to the promotion of intelligent design in public schools. The institute sent Buckingham a video entitled *Icons of Evolution* (from the title of a popular ID book by author Jonathan Wells), which he arranged to be shown to the teachers to "educate" them about the real nature of ID.[5] Interestingly, two lawyers from the Discovery Institute also made a presentation to the board, obviously a prelude to the sure legal challenge that would ensue if the board kept pursuing this clear breach of church-state separation.

Between the summer of 2003 and that of 2004, the board shifted to a delay tactic to force the teachers' hand, refusing to approve the purchase of a standard textbook, *Biology* by Kenneth Miller and Joseph Levine (Miller later testified as an expert for the plaintiffs at the trial). Despite the fact that the book had been approved by teachers (i.e., those people who actually know something about biology) and by the administration, Buckingham felt that it covered evolution too thoroughly and did not give creationism a fair shake.[6] This is like complaining that a textbook in astronomy is too focused on the Copernican theory of the structure of the solar system and unfairly neglects the possibility that the Flying Spaghetti Monster is really pulling each planet's strings, unseen by the deluded scientists.

In June 2004, some members of the board went even more clearly on the offensive, with Buckingham stating that *Biology* was "laced with Darwinism," a comment that followed another one made previously by the same character (as reported in the trial's decision) to the effect that the separation of church and state is "a myth" and, at any rate, something that he, Buckingham, personally doesn't support. One has to wonder at the size of the egos of these people who would readily put their own ideological opinions above the constitutional guarantees of an entire nation. But such is the nature of the evolution-creation debate. And speaking of church-state separation, it turns out that at the same board meeting, Buckingham's wife gave a long speech[7] (beyond the standard allotted time) during which she said that "evolution teaches nothing but lies" and asked the audience how one could teach anything but the Bible to their kids, ending up with exhorting people to become born-again Christians. Her husband came to her aid by challenging the onlookers to trace their ancestry to monkeys (nothing could be easier, as a matter of scientific fact), accusing judges in previous trials against creationism of "taking away the rights of

Christians," and ending with a call to stand up for Jesus. The point here, of course, is not that Mr. or Mrs. Buckingham or anyone else doesn't have the right to their religious opinions, or that they cannot express them in public (though a school board meeting hardly seems like the most appropriate venue). But all of this made it into the official documents of the trial because it established a crucial point for the judge: the clear religious motivations of the board in passing its resolution, and therefore the untenability of the board's legal position that its actions were meant in a secular spirit to further education and critical thinking among local students.

Of course, Buckingham himself was never concerned with respecting either other people's opinions or upholding minimum educational standards. The court proceedings relate an earlier episode, in 2002, when a mural about evolution (put up by students as a class project) was taken down and burned. When Buckingham was asked, two years later, if he knew anything about the episode, he replied, "I gleefully watched it burn."[8] This has nothing to do with Christianity or religion; it is simply the ugly face of ideological bigotry.

But let's proceed with some further background before we get to the actual trial. During the summer of 2004, Buckingham found out about the existence of an ID book called *Of Pandas and People*, produced by the ironically named Foundation for Thought and Ethics, an organization classified by the Internal Revenue Service as a public charity, but which describes itself as religious in nature in its articles of incorporation and tax returns. This book was revealed at trial to be nothing but blunt creationist propaganda, and a witness for the plaintiffs, philosopher Barbara Forrest, was able to show Judge Jones that previous drafts of the book included the word "creationism" in all the places where now it had "design." Moreover, the changes were made around 1987, immediately after the Supreme Court established that creationism is not science, but religion, and therefore cannot be taught in public schools. Coincidence or (mischievous) intelligent design?

Buckingham had the brilliant idea of forcing the Dover school system to adopt *Pandas* as a reference book, but how could he convince the teachers to go along? Simple: by blackmail. He held the approval of *Biology* hostage, clearly stating that he would allow the latter book to be purchased by the school system only if *Pandas* were adopted as well. This isn't something made up by the liberal press: Buckingham openly stated at the trial that "if I didn't get this book, the district would not get the biology book."[9] Thoughtful and ethical indeed.

All of this was not happening without some resistance and great discussion in the community. In fact, in August 2004, the board solicitor warned the Dover School Board of the possible legal consequences of their actions, stating that "my main concern at the moment, is that even if use of the text [*Pandas*] is purely voluntary, this may still make it very difficult to win a case."[10] Those words revealed themselves to be prophetic on 20 December 2005, when Judge Jones handed out an unquestioningly negative verdict for the school board.

Of course, humans being what they are, the board forged ahead, eventually forcing the purchase of *Pandas* on the teachers despite strong opposition by the latter. And yet the strangeness of the case certainly doesn't end here. In October 2004, sixty copies of *Pandas* were mysteriously donated to the school system, and at trial it emerged that both Bonsell and Buckingham had attempted to hide the source of funding, which turned out to be a call for donations that Buckingham himself did at Harmony Grove Community Church, where he attended Sunday services. The judge noticed that this meant that both these questionable characters actually lied under oath at a deposition they made on 3 January 2005, in preparation for trial, when they denied having any knowledge of the source of funding for the purchase of the sixty copies of *Pandas*.[11] I guess a lie doesn't send you to Hell if it is for the Good Cause (as Martin Luther himself famously said).

Also in October 2004, the Curriculum Committee drafted the statement that eventually led to the lawsuit. There were several irregularities and rather unusual happenings in the way the board acted on the recommendation of the committee, including the fact that the teachers—that is, those who actually know about biology—were simply not called to comment on various drafts of the resolution. Even more irritatingly, several board members who voted for the resolution, beginning with Buckingham, later frankly admitted that they had no idea what intelligent design theory actually says, nor were they in any position to knowledgeably comment on the adoption of *Pandas* as a reference book. As Judge Jones points out in the final decision for the trial,[12] it also didn't occur to anybody on the board to consult serious scientific or educational organizations, such as the National Academy of Science or the National Association of Biology Teachers. Only the ID think tank the Discovery Institute was asked for an opinion.

Here is the complete statement that the board decided should be read by teachers to students of the Dover district:

> The Pennsylvania Academic Standards require students to learn about Darwin's Theory of Evolution and eventually to take a standardized test of which evolution is a part.
>
> Because Darwin's Theory is a theory, it continues to be tested as new evidence is discovered. The Theory is not a fact. Gaps in the Theory exist for which there is no evidence. A theory is defined as a well-tested explanation that unifies a broad range of observations.
>
> Intelligent Design is an explanation of the origin of life that differs from Darwin's view. The reference book, *Of Pandas and People*, is available for students who might be interested in gaining an understanding of what Intelligent Design actually involves.
>
> With respect to any theory, students are encouraged to keep an open mind. The school leaves the discussion of the Origins of Life to individual students and their families. As a Standards-driven district, class instruction focuses upon preparing students to achieve proficiency on Standards-based assessments.[13]

Before we proceed further, let us pause for a moment and take a closer look at this statement. The first paragraph singles out the theory of evolution for particular attention by the students. This is an inappropriate signal that there may be something especially wrong with this and no other scientific theory. Surely the Pennsylvania Academic Standards establish a variety of benchmarks for the students of that state in order to demonstrate academic proficiency, but the board clearly wanted to set evolution aside for special consideration. Since evolutionary theory is taken (by scientists) to be as good a scientific theory as, say, continental drift or quantum mechanics, this decision by the board was obviously motivated by religious views, which is one of the several reasons that led Judge Jones to rule against the school board in the end.

The second paragraph is beautiful because it is at once largely correct and deeply misleading. First, as I mentioned at the onset of this chapter, it is certainly true that "Darwin's theory is a theory," in the same way that it is true that my chair is a chair, that is, by definition (notice, incidentally, that just like modern mechanical physics is not "Newton's theory," the modern theory of evolution is not "Darwin's theory," but a more advanced and sophisticated theory known as the Modern Synthesis). Second, the theory is not a fact, again, by definition. Scientific theories are just what the board states they are: explanations meant to unify a broad range of observations. Facts are observations that the theory attempts to make sense of, which is why the standard creationist refrain that "evolution is a theory, not a fact" is at the same time (almost) correct and yet entirely irrelevant. Moreover, the business about gaps in the theory "for which there is no evidence" betrays a profound misunderstanding of the nature of scientific theories: every theory has "gaps," meaning sets of facts that are not (currently) explained by the theory or, conversely, makes hypothetical statements that are not (currently) supported by empirical evidence. This is true for evolution, quantum mechanics, general relativity, or whatever. Theories are human constructs meant to guide our understanding of nature, and as such have always been and will always be "incomplete." Once one realizes that this is a normal condition of science, it will not come as a surprise that the theory of evolution has "gaps." Furthermore, there is no reason for students to focus on those particular gaps in contrast to any other lacuna of any other scientific theory.

The third paragraph contrasts intelligent design with "Darwin's view" (again, the proper contrast should be with the modern theory of evolution, not with Darwin's original views—science makes progress with time). It states that ID is an alternative explanation and that students can find more about it in the "reference text" *Of Pandas and People*. But, in fact, ID is not a scientific theory at all because there is no empirical observation that can possibly contradict it. Anything we observe in nature could, in principle, be attributed to an unspecified intelligent designer who works in mysterious ways. As long as we do not venture to make hypotheses about who the designer is and why and how she operates, there are no empirical constraints on the "theory" at all. Anything goes, and therefore nothing holds, because a theory that "explains" everything really explains nothing.

Let me elaborate on this point, because it is crucial to our discussion. ID proponents such as William Dembski are fond of drawing a parallel between their position and the sort of conclusion one would reach while looking at Mount Rushmore in South Dakota. This mountain, as is well known, was carved between 1927 and 1941 (before the modern environmental preservation movement took hold) with the facial features of four American presidents: Washington, Jefferson, Theodore Roosevelt, and Lincoln. Now, ID proponents say, it doesn't take a rocket scientist (or, apparently, an evolutionary biologist) to figure out that the faces were intelligently designed and were not the result of a natural process. They pompously call this the "design inference."

Indeed, to deny the validity of such an inference would be ludicrous. But the reason such a conclusion is inescapable is because we know a lot about the intelligent designers themselves: we know that they are human, how they managed to carve the mountain, and why they did it. It is not (just) the complexity or the features of the faces that allow the design inference, it is the fact that the inference is a hypothesis made testable by our additional knowledge concerning human monuments. We have absolutely no such information in the case of the alleged intelligent designer of biological organisms; moreover, while there is no alternative, non-natural explanation available for the faces on Mount Rushmore,[14] there is a perfectly viable scientific theory that tells us how living organisms change over time: through a series of mechanisms that involve the generation of random mutations and nonrandom survival and reproduction of those organisms that happen to be better adapted to the local environmental conditions. There is much more to the modern scientific theory of evolution, but the contrast between the two is sufficient to make clear that ID is not, in any meaningful sense of the word, a theory of anything.

The last paragraph in the board's statement is a masterpiece of deception. To begin with, no scientist or educator would ever argue against teaching critical thinking to students, and therefore naturally we would like students to "keep an open mind" about anything. But, again, why was the theory of evolution singled out, if not for religious purposes? Should Dover's students not be equally open minded about, say, general relativity or the germ theory or, for that matter, astrology (or even religion)? Open-mindedness means that one does not accept a certain conclusion as dogma, but only tentatively and in proportion to the available evidence. By that standard, the scientific theory of evolution is what an open mind *should* accept, since it is overwhelmingly supported by the available empirical evidence. Could it be that the theory will be superseded by another one in the future? Yes, and in fact that is what happens to most scientific theories because science is a progressive enterprise (in the sense that it improves its own products) and it is part of the job description of a scientist to keep an open mind and be willing to abandon a theory that no longer works. This has happened already to several other theories of evolution (for example the one advanced before Darwin by the French biologist Jean-Baptiste de Lamarck) as well as to Newtonian mechanics and even the early version of the Copernican theory (Copernicus was right that it is the earth that rotates around the sun and not vice versa, but he was wrong in thinking that such an orbit was circular; it is elliptical).

Finally, the statement about the fact that students will not be taught about the origin of life is downright puzzling. The origin of life is a scientific question and as such should be discussed in science classes. Of course students are free to also discuss it with their families, especially when it comes to its perceived metaphysical and religious implications. But the role of education is to provide students with both the best available understanding of a given problem to date and the reasoning tools to think independently about the problem itself. Not teaching a subject—no matter how controversial—is not good education; it is simply subordinating education to an ideological agenda. Which is exactly what got a minority of board members and the teachers' body to publicly distance themselves from the board's peculiar decisions and modus operandi.

Let us now resume our chronicle of the bizarre events at Dover, as reported in the trial proceedings, to help us reflect on the comprehensive picture—not to mention its ugly emotional underpinning—of the so-called evolution-creation controversy that emerges from them. After the board passed its inane resolution, opposition became more vocal, beginning with some of the board members themselves. For example, Casey Brown wrote a letter of resignation from the board that, in part, states: "Our opinions [on the board] are no longer valued or listened to. . . . I myself have been twice asked within the past year if I was 'born again.' No one has, or should have the right, to ask that of a fellow board member."[15] Indeed, and note that this statement does not come from an avowed atheist (although if it did it shouldn't make any difference), but from someone who simply happened to believe in the separation of church and state. No matter: Buckingham called Brown an atheist, and Bonsell told her—according to the trial's documents[16]—that she would go to hell. So much for Christian compassion and open mindedness.

This was not an isolated incident either. The trial proceedings show that the second board member to resign, Noel Wenrich, explained in a letter to the board that "I was referred to as unpatriotic, and my religious beliefs were questioned,"[17] even though Wenrich actually served in the U.S. armed forces for eleven years and is, in fact, a Christian. This is not a matter of religion versus science or even versus atheism; it is a simple matter of religious bigotry and intolerance on the part of a particular group of Christian fundamentalists.

Despite all of this, the board approved the disclaimer about evolution that teachers were supposed to read to their students, except that the teachers refused to cooperate. They sent a letter to the board that said that the board's request "violates our responsibilities as professional educators" and that "to refer the students to *Of Pandas and People* as if it is a scientific resource breaches [the teachers'] ethical obligation to provide them with scientific knowledge that is supported by recognized scientific proof or theory."[18] Strong stuff, which clearly set the stage for the ultimate confrontation in court.

Things were not much calmer in the surrounding community either. Judge Jones actually admitted as evidence at trial all the letters to the editor on the matter sent to two local newspapers, the *York Dispatch* and the *York Daily Record*, as well as editorials about the board's actions published in the same papers. The numbers are staggering: from 1 June 2004 to 1 September 2005 a total of 62 editorials and 225 letters appeared in the *Dispatch* and the *Record*! This was relevant, as Judge Jones explained,[19] not to establish the validity of either position, but rather to confirm that a reasonable member of the community would have been justified in perceiving the actions of the board as an attempt to impose a particular religious perspective in a public school, in violation of the First Amendment. The statistics speak for themselves: of 139 letters published by the *Record*, 86 clearly framed the issue in religious terms (pro or against); similarly for 28 of the 43 editorials published by that paper. In the case of the *Dispatch*, the analogous figures are 60 out of 86 letters, and 17 out of 19 editorials. Can there be any reasonable doubt that this is not a matter of teaching science, but a straightforward infringement of the establishment clause?

Predictably, the community was bitterly divided by the board's stance, as was testified by Joel Leib, a local resident with multigenerational ties to the area: "[the Board's actions have] driven a wedge where there hasn't been a wedge before. People are afraid to talk to people for fear . . . They're afraid to talk to me because I'm on the wrong side of the fence." Yet, as we have already seen, and despite all the obvious evidence to the contrary, the defendants at the Dover trial had the audacity to present their motivations as "secular" and their actions as aimed only at furthering the education and critical thinking of the students. As Judge Jones (again, remember, a conservative judge appointed by the second President Bush) aptly put it: "To assert a secular purpose against this backdrop is ludicrous."[20]

A Long History of Legal Battles and Shifting Strategies

The Dover trial is just one of the most recent (though historically crucial) examples of the ongoing legal war between religious zealots and defenders of church-state separation in the United States. Judge Jones himself pointed out in his decision at Dover that such history is relevant to the understanding of the controversy because it clearly shows the shifting tactics employed by creationists, thereby plainly exposing their nonsecular, and not really well-hidden, motives.

The opening salvo, as is well known, was the infamous Scopes trial of 1925, which took place in Dayton, Tennessee.[71] John Scopes, a substitute biology teacher, was arrested on the grounds that he was teaching the illegal doctrine of evolution (it is not clear whether he actually did or just pretended to so that the trial could take place and the town of Dayton could be "put on the map," as local community leaders were hoping). Scopes was convicted—the only time creationists actually won a major legal challenge—despite the participation in the trial of renowned defense lawyer Clarence Darrow and of the American Civil Liberties Union (then a fledgling organization at one of its first high-profile cases) on the side of Scopes. On the prosecution side was three-time Democratic presidential candidate William Jennings Bryan, after whom the local Bryan College is still named.

The Scopes trial was the stuff of Hollywood movies, and in fact one was made in 1960, followed by three more movies made for TV. The 1960 version features Spencer Tracy playing Darrow, Fredric March in the role of Bryan, and an unusual dramatic role for Gene Kelly, playing a journalist reminiscent of *Baltimore Sun*'s H. L. Mencken, who in real life filed scathing commentaries on the trial and the local culture for the *Philadelphia Inquirer* (the movie, and the play from which it was adapted, *Inherit the Wind*, was actually meant as a criticism of the abuses of the McCarthy era, which occurred several decades after the Dayton trial). Scopes's conviction was overturned on a technicality, and the state of Tennessee did not dare retry the ill-conceived case, fearful of even more bad publicity and Mencken-style barbs (a classic example of which is "It is hard for the ape to believe he descended from man"). The result, however, was that the law prohibiting the teaching of evolution in Tennessee actually stayed on the books for decades until it was finally repealed in 1967.

But why did the Scopes trial occur in 1925, almost seventy years after the publication of Darwin's *Origin of Species*? Partly because the early twentieth century had seen a religious revival in the United States with the publication of a series of books called "The Fundamentals" (from which the word "fundamentalism" derives), published to bring people back to the "original" word of the Bible (of course, there is no longer such a thing as the original manuscript of the Bible, and we have to rely on the earliest surviving manuscripts in Hebrew and Aramaic, together with an influential translation in ancient Greek). Indeed, Bryan was a populist politician who was profoundly disturbed not by the scientific theory of Darwin itself (about which, arguably, he understood little) but by decidedly unscientific "social Darwinism" ideas, used by some scientists and politicians to advocate inhumane treatment of the poor based on a cynical misunderstanding of the science of evolution.

The approach used by fundamentalists during the early stages of their legal war was simply to attempt to prohibit the teaching of evolution. That tactic came to an abrupt end in 1968, when the Supreme Court struck down an Arkansas statute that aimed at making the teaching of evolution illegal. The Supreme Court is the last appeal in legal matters in the United States, so once it declares something unconstitutional, there is no other possible recourse—except waiting for the makeup of the court to change in a favorable manner, or changing the Constitution itself. The 1968 decision marked the first major shift in tactics on the part of creationists: since banning the teaching of evolution was no longer feasible, creationists came up with the idea of a "balanced treatment," that is, of teaching both evolution and creationism in the name of presenting students with a choice about competing "theories."

That shift in tactics, however, proved in turn to be short-lived: in 1987 the Supreme Court again intervened in the controversy, this time declaring any "balanced treatment" approach, and indeed the very teaching of creationism, in violation of the First Amendment to the U.S. Constitution. Understandably, this caused significant disarray among creationists, who were seemingly left with nowhere to go—until they invented intelligent design theory. It is not by chance that the ID movement got off the ground in the early 1990s, largely as a result of the public advocacy of retired Berkeley law professor Phillip Johnson. If creationism didn't make it under the pretentious name of "creation science," then perhaps it would under the more ambiguous term "intelligent design theory." It is this continuously shifting series of tactics in the single-minded pursuit of state establishment of religion that led philosopher Barbara Forrest and biologist Paul Gross to entitle their book on intelligent design *Creationism's Trojan Horse*.[22]

This is also why Judge Jones's decision is so important: although it does not carry the weight of a Supreme Court decision (and, in fact, we can fully expect creationists to push the issue again within the next few years, until it does reach the highest court in the land), it marks the beginning of the end for the third round of creationist legal strategies. It seems like they are truly running out of options now, and pathetic attempts to rename ID along the lines of "sudden emergence theory" (a bizarre term that has surfaced here and there in the aftermath of Dover) surely are not going to cut it.

This complex legal history is why we need to understand the criteria applied by Judge Jones to reach his decision at Dover: the endorsement test and the so-called Lemon test of the First Amendment[23] (after the case of *Lemon v. Kurtzman* [1971], not the fruit). The endorsement test was explained by the Supreme Court: "School sponsorship of a religious message is impermissible because it sends the ancillary message to members of the audience who are nonadherents that they are outsiders, not full members of the political community, and an accompanying message to adherents that they are insiders, favored members of the political community."[24] Judge Jones clearly concluded that this was indeed the case at Dover, and it is hard to find a more blatant example of it than the abovementioned accusation of atheism and lack of patriotism hurled by board members to the dissenting minority. One has to remember, of course, that the establishment clause works in both directions: it not only prohibits the state from endorsing (any) religion, it also forbids the state interference with religious practice. Consequently, Mr. Bonsell and Mr. Buckingham are perfectly free to advocate any view they may hold on the origin of life and its history, both in their churches and in the public square, but they are barred from doing so in their capacity as public officers.

Let us now turn to the Lemon test. First, this has been considered necessary by the courts only if the endorsement test fails: that is, if one can convincingly show that a statute or action directly endorses a particular religious viewpoint, this is sufficient to rule against such statute or action, and no Lemon test is necessary. The latter becomes relevant in cases where a breach of the First Amendment could not be reasonably demonstrated under the endorsement test alone. Lemon hinges on precise standards regarding the purpose and the effect of the statute or action being challenged. Judge Jones summarized the Lemon (as articulated by the Supreme Court) in this way: "under the Lemon test, a government-sponsored message violates the Establishment Clause of the First Amendment if: 1) it does not have a secular purpose; 2) its principal or primary effect advances or inhibits religion; or 3) it creates an excessive entanglement of government with religion."[25] Notice that the first point regards the purpose, the second and third, the effect of a given statute or action by the government.

It is the Lemon test that explains, for example, why Judge Jones was interested in the newspaper editorials and letters published about the controversy: they clearly revealed that the actions of the board were not motivated by secular aims, fulfilling the "purpose" component of the Lemon test. The same literature, as well as the letters of resignation written by some minority members of the board, and especially the letter written by the teachers to the board, all make clear that the board's maneuvers did create an excessive entanglement of government and religion and that the primary result of such maneuvers was, in fact, to advance a religious viewpoint. This equally clearly fulfilled the "effect" part of the Lemon test.

So, Judge Jones—unmistakably conscious of the importance of the Dover case—played it particularly safe and made sure that his decision was based on both the establishment test as well as on the purpose and effect aspects of the Lemon test. The defendants failed on all counts, leaving no doubt that this was as clear a case of violation of church-state separation as one is likely to find in the annals of jurisdiction.

But Is It Science?

While we have learned much about creationism and intelligent design from a careful analysis of the Dover proceedings, the crucial question as far as this book is concerned is rather more direct: is ID science, under any reasonable definition of the term? Just because an idea has strong religious underpinnings, it doesn't necessarily make it is scientifically unsound, and one ought to pursue the two questions (infringement of state-church separation and scientific nature of the claim) separately. This is, in a sense, what Judge Jones did, and the central part of his decision, entitled "Whether ID is science,"[26] should be a must-read in any discussion of science and religion. Interestingly, Judge Jones relied not only on the expert testimony of ID proponents and of their scientific critics, but also on a usually neglected category of experts: philosophers of science. His account is an enduring testimony to how a person with no technical background, a religious believer, and a political conservative, can form a solid opinion—based on considerate understanding of expert testimony—of matters that are both intellectually sophisticated and full of emotional and political implications.

Jones clearly stated in his discussion that he does not take any position as to the truthfulness of ID itself. This is also what most scientists do: intelligent design, or other forms of creationism, may be true (although it is hard to imagine how one could ascertain it), but the controversy hinges upon the twin questions of whether ID is religion (it is) and whether it is science (it isn't). Nonscientific claims may be true and still not qualify as science. For example, it is true that I am writing this while in a hotel in Vienna during a work-related trip; moreover, it is a truth that is empirically ascertainable, since there is a record in terms of eyewitnesses (the colleagues to whom I gave a talk last night), hotel bills, airline tickets, and credit card receipts. However, this particular truth is not "scientific" in any meaningful sense of the term because it isn't part of any general theory of how the natural world works.

Judge Jones concluded that ID fails as a science on three grounds: "(1) ID violates the centuries-old ground rules of science by invoking and permitting supernatural causation; (2) the argument of irreducible complexity, central to ID, employs the same flawed and illogical contrived dualism that doomed creation science in the 1980's; and (3) ID's negative attacks on evolution have been refuted by the scientific community."[27] It is crucial to understand these three points because they apply to the creation-evolution controversy in general and because they represent an excellent example of the distinction between science and pseudoscience.

First, then, let us examine what Jones meant when he said that ID (and, in fact, any form of creationism) violates the ground rules of science. As we shall see in the next chapter, one can trace elements of the scientific approach to ancient Greece, and in particular to Aristotle, although it is generally agreed that modern science really begins with Francis Bacon and his articulation of the inductive method. But even back in Aristotle's time, a fundamental assumption of doing things scientifically is that the supernatural is out: no explanations that invoke non-natural causes are allowed. It is not difficult to find clear statements to such effect in the ancient literature. For example, Epicurus (341–270 BCE), who was nineteen when Aristotle died, wrote in a letter to his friend Pythocles: "The divine should not be introduced in any way into these considerations [theories about nature] . . . If this is not done, all speculation [i.e., theorizing] on the causes of heavenly phenomena will be in vain."[28]

But wait, is scientific practice then based on unquestioned assumptions? How can it be scientific to rule something out (or in) a priori, without empirical evidence? The answer is that this aspect of the scientific method is not, in fact, scientific: it is a philosophical position. As Robert Pennock and Barbara Forrest—two philosophers of science who testified at Dover—have explained, this assumption is known as "methodological naturalism." It is worth bearing with me a little here, because the common misunderstanding of this point is a major reason why so many people consider ID the "reasonable" compromise that it most certainly is not. Bill Demsbki, a leading proponent of ID, has repeatedly claimed that he is trying to broaden the current concept of science, to get science to move beyond naturalism to take into account the possibility of supernatural explanations. The debate hinges upon three crucial terms: "naturalism," and two adjectives that can be attached to it: "methodological" and "philosophical."

Naturalism, in this context, is simply the idea that the world works according to natural laws and processes, but it comes in two varieties: methodological and philosophical. A philosophical naturalist is, essentially, an atheist, or someone who believes—as a result of philosophical (not scientific) considerations—that there really is no supernatural at all. This is a minority (though perfectly reasonable) position in the population at large, and many scientists do not espouse it either. But how can one be a scientist and reject philosophical naturalism? By being a methodological naturalist, of course! Methodological naturalism is at the core of science because it doesn't commit a scientist to atheism; it simply says that—since science cannot possibly investigate the supernatural—the supernatural, if it exists, cannot factor into scientific explanations of how the world works. This is not at all the same as saying that the supernatural doesn't exist; it is simply, in a sense, to admit the limitations of science in being able to deal only with natural causes and empirical evidence. At the same time, it frees science from any close tie with religion and allows scientists to pursue their work independently of their private religious beliefs.

Now, this is not a particularly strange position, contrary to what Dembski and company would argue. Indeed, almost everyone is a methodological naturalist at heart, regardless of his religious creed, and all of us certainly behave that way in everyday life. Consider what might happen the next time your car doesn't start. Very likely your first thought isn't going to be that some supernatural agent wishes to interfere with your plan to get to the grocery store. Instead, you will look at the gas gauge to make sure that there is enough fuel in the tank. If that checks out, you'll proceed with a series of additional controls, such as making sure the battery is operating normally, and so on. In other words, you are assuming—without any direct evidence—that there is a natural explanation for why your car is not running. Let us push the analogy a bit further and suppose that you go through all the standard checks and still can't find anything wrong with the car. Again, I bet you will not take out your cell phone (most definitely not a supernatural device, though clearly intelligently designed!) and call your pastor, rabbi, or imam. You will call your mechanic, sticking with your methodologically naturalistic assumption about cars in general. Moreover, suppose your mechanic calls you back a week later and says that he couldn't identify the cause of the malfunction and that the car still won't run. Most likely you will persist (despite the complete lack of evidence) in assuming that there must be a mechanical (i.e., natural) explanation and that your mechanic either isn't competent enough or had insufficient information to find it. You may shrug your shoulders and go to the nearest car dealer to buy another car. Even if you are a religious believer, throughout this process you behaved just like a scientist, and for the same exact reason: the only explanation you can possibly act upon is a natural one. If God truly didn't want your car to start, then there is nothing you could do about it anyway, and no way you could prove (or disprove) it either.

Creationists are fond of pointing out that scientists *assume* that there must have been a natural cause to, say, the origin of life, without really knowing that this is indeed the case. It should be clear now in what sense scientists make this assumption and why this is in fact a valid and reasonable way to go about things. The assumption is not philosophical, it is methodological: scientists do not *conclude* that there was a natural cause for life's origin and then go home satisfied. They work under the provisional contention that there had to be a natural cause because, just like in the case of your mechanic, that's the only way they can hope to find an explanation. They may succeed or fail, but neither outcome will actually say anything at all about the existence or workings of God.

All of this is also why when Dembski and other ID proponents say that they wish for a "broadened" and more complete science, they are talking nonsense. Supernatural explanations aren't explanations at all, because one doesn't know why and how God decided to do what he decided to do (after all, he is God!),[29] which means that any so-called explanation that invokes the supernatural turns out to be nothing more than an elaborate admission of ignorance: instead of simply saying that we have no clue about why or how something happened (the car isn't working) we increase our self-importance by concluding that God personally got involved into the matter.

Judge Jones, therefore, rightly concluded that ID would change the very definition of science and that this change is unwarranted. Defense witness Scott Minnich, a microbiologist, admitted: "for ID to be considered science, the ground rules of science have to be broadened to allow consideration of supernatural forces,"[30] thereby violating methodological naturalism. Perhaps worst of all for the defense, plaintiff attorney Eric Rothschild got biochemist Michael Behe (a prominent ID supporter) to state that his "broadened definition of science . . . would also embrace astrology."[31] In fact, an argument can be made that astrology is not quite as bad as ID because the former only postulates natural forces unknown to science, not necessarily supernatural ones, but the comparison is damaging enough that Behe will regret having made it on record for the rest of his life.

Let us now consider the second major conclusion reached by Jones: that ID's central argument, known as "irreducible complexity," is simply an example of the same "contrived dualism" that the courts found in the 1980s was a fatal flaw of standard creationism. We need therefore to understand both what irreducible complexity is supposed to be and why contrived dualism is a logical fallacy.

"Irreducible complexity" is a term introduced by the above-mentioned Behe in his influential *Darwin's Black Box* and expanded and elaborated upon by Dembski in several books and articles.[32] However, as Behe himself admitted at Dover, the basic idea goes back to the writings of Rev. William Paley and his *Natural Theology* of 1802.[33] Paley is famous for having proposed the analogy between the intelligent designer and a watchmaker as the major justification for inferring intelligent design in nature. He asked his readers to imagine themselves walking on a beach and looking at scattered pebbles. Surely they would not be surprised to find them, and they would rightly assume that natural processes are the cause of the pebbles' sizes and shapes. However, Paley says, imagine then that you stumble upon a watch. You immediately realize that this is not a natural object, that it could not possibly have been the result of a natural process. You (correctly) infer that it must have been "intelligently designed" by a watchmaker.

If this sounds familiar, it is because Paley's example is exactly analogous to Dembski's "design inference"[34] applied to the figures sculpted on Mount Rushmore: in both cases we do in fact conclude that there was an intelligent agency, and not a natural process, at work. But why? The real answer should now be obvious to the reader: because we know lots of things about watches and sculptures, and in particular about watchmakers and sculptors. We know how they work and why, which is how we can recognize the products of their handiwork. But this, clearly, won't do for proponents of intelligent design because they don't want to admit anything about the designer and why or how she works (otherwise it would be obvious that they are talking about God). Hence Behe's subterfuge of introducing the rather ambiguous concept of irreducible complexity.[35]

Behe says that what watches, the faces on Mount Rushmore, and some biological structures have in common is that they could not possibly have come about by natural processes because they are made of a high number of parts that are supposed (designed) to work together: take any component out, and the whole falls apart. His favorite example of an irreducibly complex structure is the mouse trap[36] (hence the quote at the beginning of this chapter): exclude any part, and the trap stops working. Of course, the problem with the analogy is that mouse traps *are* unquestionably intelligently designed, and—again—we know much about the designers, their methods, and their intentions. (Moreover, and rather amusingly, John McDonald at the University of Delaware has actually produced a video showing that mouse traps are not, in fact, irreducibly complex, as one can obtain simpler and simpler mouse traps by taking out one part at a time while the trap continues to work—all the way down to just one component.[37] The only problem is that they become increasingly inefficient at catching mice!)

Now, since nature does not have forethought, then the only way to get irreducible complexity, according to Behe, is by intelligent design. To put it another way: just as half a watch isn't any good at telling time, half an eye (Paley's original parallel between manmade artifacts and biological structures) is not good either. Except, of course, that it is. It turns out that biologists have figured out, over the last several decades, how complex eyes can and did evolve: we now have evidence from the fossil record, from currently living organisms, and even from mathematical models[38] that half an eye is, well, half as good, but still a heck of a lot better than not seeing at all!

Behe knows about research on the evolution of the eye, which is why he doesn't use that old workhorse of Paley-style intelligent design. Instead, Behe retreats to the molecular level, which is more esoteric and much less familiar to the general public, suggesting that complex molecular structures and processes, from the bacterial flagellum to the blood-clotting biochemical cascade to the immune system, cannot possibly be explained by evolutionary theory. More on these examples in a moment, when we will take up Judge Jones's third and last conclusion about ID.

First, however, it is important to realize why the basic logic of irreducible complexity is flawed, regardless of specific examples. Essentially, Behe's and Dembski's idea is to prove a negative (evolution cannot produce a complex structure, particularly by means of natural selection) by arguing that there could not have been simpler intermediate structures (because of the "irreducibility" attribute). Since any naturalistic theory of evolution requires that complexity emerges gradually through intermediate, less complex steps, then pronto, we have "demonstrated" that evolution couldn't possibly happen, from which conclusion we then embrace the only other choice: (supernatural) intelligent design. Not so fast, says Judge Jones.

The judge pointed out that this sort of "reasoning" (to be very charitable) is the same "contrived dualism" that standard creationism suffers from. Contrived dualism is a logical fallacy wherein one is forced to make a choice between two alternatives, while in fact there are other possibilities that are being conveniently ignored. For example, a real estate agent who wants to sell you a particular house may try to present you with an "either you buy this one (because it is such an incredibly good deal) or you will not be able to find anything else" sort of argument. But of course in reality you do have the choice of looking at other houses in the same neighborhood or even of going to a different area of town (or of changing real estate agents). Analogously, the argument from irreducible complexity sets up a false dichotomy: either evolutionary theory can explain everything or ID is true. But a moment's reflection shows that there are several other possibilities: since all scientific theories are incomplete, any one of them may be unable to explain something at any given point in time, but not necessarily in the future. Or maybe the scientific theory in question really will be discarded eventually (many theories have suffered that fate in the past), but a better theory will be produced without having to accept ID as the default alternative.

Moreover, evolutionary theory actually does have an explanation for apparently irreducibly complex structures: it involves the concept of "exaptation" (as opposed to adaptation), and it states that some structures evolve gradually at the same time as they are changing their function. Exaptation is a bit like reusing old stuff you have in your basement for new purposes, maybe while at the same time altering the object to better suit its new role (indeed, biologist and Nobel winner Francois Jacob made an analogy between evolution and "bricoleur," the French word for tinkerer[39]). For example, you may have an old TV stand with drawers, but you have now decided to spend more time reading books than watching TV. So you get rid of the television, and you are left with the stand; it occurs to you that it would actually serve as a decent chest of drawers for your bedroom, especially after you repaint it. Notice that the old structure (the TV stand) is not ideally suited to the new function (chest of drawers), but it will do well enough, and over time you can modify it to improve its new role (e.g., by painting it).

There are several documented examples of this type of process in biology, one of the most spectacular being the slow transition and modification of some bones from the jaw of ancient reptiles to the middle ear of their mammalian descendants. This sort of evolution is made possible by the fact that living organisms are redundantly complex, to use a term proposed by philosopher Niall Shanks and biologist K. H. Joplin.[40] Unlike irreducible complexity, redundant complexity is a situation in which many parts carry out many functions, and the system is messy enough (because it wasn't intelligently designed) that some parts can cease to function for a particular task and take on another one over time, without the system as a whole being disrupted in the process. This is what we would expect from complexity arising from a natural historical process, rather than originating from a superintelligent engineer.

We finally get to Judge Jones's last reason for why ID is not science: its negative attacks on evolution have actually successfully been rebutted by biologists, leaving ID with nothing to show for itself, either in the way of positive contributions to knowledge or as a serious critique of an accepted scientific theory. The trial transcripts provide fascinating in-depth discussions of each of the three major objections raised to the explanatory power of evolutionary theory by ID supporter Michael Behe: the bacterial flagellum, the blood-clotting cascade, and the immune system. Behe's argument is that the structure of the flagellum (essentially a rotating whip used by some bacteria to move through fluids), the biochemical reactions of the blood-clotting cascade (which stop the bleeding after a skin cut in animals), and the details of the immune system's response are "textbook" examples of irreducible complexity and could not possibly have evolved through intermediate forms. Behe contends that a flagellum with some parts missing could not do its work at all, that skipping even one biochemical reaction in the blood-clotting cascade would lead to certain death, and that biologists have no idea of how the immune system could have originated from simple beginnings. How, then, could these systems possibly have evolved from earlier versions?

It turns out that Behe is simply wrong, and (possibly willfully) ignorant of the relevant literature. Plaintiff expert Kenneth Miller, a biochemist at Brown University (and the coauthor of the widely acclaimed biology textbook that the Dover School Board at one point refused to purchase), explained to the judge that there are, in fact, intermediate structures that preceded the evolution of the flagellum.[41] For example, scientists have discovered that some bacteria use a subset of the same proteins in what is known as the type III secretion system. This structure is deployed by the bacterial cell not as a flagellum, but as an injection pump. The point is that here we have a precursor to the flagellum, parts of which could then have been used by bacteria to perform a different function, just like your former TV stand that becomes a chest of drawers.

Miller also explained in court about the blood-clotting cascade: in direct contradiction to Behe's assertions, the scientific literature shows that there are plenty of intermediates of the cascade.[42] For example, dolphins and whales can clot despite missing an entire part of the biochemical mechanism found in other mammals. This is what the theory of evolution predicts we should observe, and it cannot be explained by ID.

Finally, Miller also brought to trial evidence from peer-reviewed scientific publications presenting hypotheses and empirical data on the evolution of the immune system.[43] A total of fifty-nine papers, nine books, and even several immunology textbooks were presented to Behe at trial in an attempt to make him admit that he was simply wrong about the alleged irreducible complexity of the immune system. His answer was that the evidence was "not good enough," though he did not elaborate on what could possibly be sufficient to make him acknowledge his mistake.

Behe was perhaps the most disastrous player for the ID camp at the Dover trial. Not only did he stubbornly refuse to admit that he was wrong when faced with overwhelming evidence of his own intellectual arrogance and ignorance of the literature; not only, as we saw earlier, has he admitted that by his criterion of science, intelligent design is on the same footing as astrology; but he made yet another monstrous blunder when questioned about design inference, the centerpiece of the ID position. As we have seen, this is the idea that when one observes certain artifacts—like a watch or the presidents' faces on Mount Rushmore—one can reasonably infer the existence of an intelligent designer who made those artifacts. As philosopher David Hume pointed out already in the eighteenth century, this argument won't cut it because of the same objections that Behe faced at Dover: human artifacts are not alive, do not reproduce, and are not subject to natural selection, so there is a disanalogy between human artifacts and living organisms. Behe's response to all of this was, and I quote, that the inference "still works in science fiction movies"![44] What is really astounding is that these buffoons actually have gotten as far as having their day in court.

Our brief analysis of intelligent design shows that it is not science because it invokes the supernatural, in violation of the methodological naturalistic approach that has characterized science since Bacon, and in fact all the way back to Aristotle. ID also makes no positive prediction of its own, but relies instead only on negative arguments, implying that if an established scientific theory fails to explain a set of phenomena, then ID wins by default. But this, as we have seen, constitutes a gross logical fallacy. The conclusion is that intelligent design is simply a religious position, not substantially different from standard creationism, and it therefore should not be taught in public school science classes (*both* because it is religion *and* because it is bad science). All of this notwithstanding, let us remember once again that science does not and cannot pronounce itself on the truth of a metaphysical idea (such as the existence of God), something best left to philosophers and theologians; *there* lies the true distinction between science and religion.

We have so far explored the complex territory that encompasses solid science ("hard" or "soft" as it may be), quasi-science, and downright pseudoscience. In the final portion of the book we will deal with the so-called culture wars between science defenders and science critics, and will eventually arrive at exploring the questions of what makes one an expert and how can we tell whether we should trust her or not. Before that, however, we need to take a detour and further solidify our understanding of science by looking at its origins and early developments, to be better equipped to grasp both its limits and potential. This is a story that began more than two and a half millennia ago and that has involved some of the most brilliant and largest intellectual egos humanity has ever produced. Someone should make a movie about it one of these days.

CHAPTER EIGHT

From Superstition to Natural Philosophy

Science is what you know. Philosophy is what you don't know.—Bertrand Russell

It is the mark of an educated mind to be able to entertain a thought without accepting it.—Aristotle

History too often gets a bad rap, and it is dangerous for an author at any point in a book to indulge in an historical detour. Nonetheless, it is also very difficult to get an idea of why we are where we are, and why we face certain problems rather than others, without having a clue about how things developed to this point. The goal of this book is to map the complex territory dividing science from pseudoscience, and it would be rather meaningless to attempt to do so without asking fundamental questions such as what science is, as we did at the outset of this volume. The pursuit of our goal, however, also requires some grasp of where science comes from and how it became the all-pervasive activity that we experience in the twenty-first century.

Rather than attempting to provide a complete chronological history of science—which would be both dull and way beyond the scope of this volume—over the next two chapters I will sketch instead some vignettes that are meant to convey a general impression of the *conceptual* history of science. We will therefore pay relatively little attention to the lives of early scientists and philosophers, as fascinating as they sometimes were, and more to how science developed as a set of ideas whose roots are lost in the mists of time. In the course of this brief exploration, we will encounter some of the fundamental notions and people that have shaped modern Western culture, as well as those that have deeply influenced human thinking in general.

From the Rivers of Babylon to the Pre-Socratics

It is often said that science began when prehistoric humans looked at the sky and wondered what it was all about. Perhaps, though it sounds to me like that was more likely the beginning of religion, and eventually philosophy, than science. After all, "what it is about" is a question concerning meaning, not how things work. In fact, what we know of the early stages of the quest for knowledge clearly indicates that it was directed toward two sorts of goals, which were often intricately connected: to solve practical problems and to figure out what the gods were up to.[1]

The ancient golden triangle where science developed had its vertices in Greece, Egypt, and Mesopotamia (today's Iraq, Iran, and Syria), but the crucial contributions of the people from these areas occurred at different times and in different fields. The Greeks, in particular, imported a significant amount of mathematical and astronomical knowledge slowly accumulated by the ancient Egyptians and Babylonians, and then ran with it to establish in a matter of a couple of centuries the foundations for what we still today recognize as philosophy and science.

Egyptians and Babylonians got an early start, with the former developing a number system by circa 3000 BCE and the latter doing the same about a millennium afterwards (things moved slowly during the early stages of human cultural evolution). The Egyptians had a calendar, though it was fixed (12 months of 30 days plus 5 bonus days at the end), and had the know-how to calculate the areas of plane geometrical figures and the volumes of a few solids (including, of course, pyramids). Babylonian math was much more developed than the Egyptian one, and their knowledge of astronomy was also much more advanced. Of particular interest, however, is what Egyptian and Babylonian scientific knowledge was *for*: these cultures were interested in agriculture, calendars, astrology, and medicine, that is, a combination of religious or mystical practices and of practical concerns. And these two types of knowledge were seamlessly intertwined. For instance, ancient medicine was based on the principle that disease was caused by evil spirits, so that the best way to deal with health problems was through the appropriate religious rituals. These rituals, in turn, were informed by the study of the intentions of the gods, accomplished through astrology, which itself was based on astronomy. Moreover, since nobody could guarantee to have interpreted the will of the gods correctly, pharmacological remedies of all sorts where prescribed and used, just in case.

This, then, seems to have been the nature of the ancient origin of science. Not a quest for knowledge of the ultimate structure of the universe (that will come shortly, as we shall see), but a hodgepodge of practices aimed fundamentally at achieving the same end: to improve humanity's lot on earth, be that through actual knowledge of nature (astronomy, agriculture) or through what we today would call superstition (astrology, spiritual healing)—though of course at the time few people could tell the difference. All of that changed radically as soon as the theater of intellectual activity shifted to Greece, beginning with the so-called pre-Socratic philosophers.[2] The pre-Socratics were an intellectually (and perhaps otherwise, who knows?) wild bunch, about which we don't know very much, since most of their ideas survive through fragments of manuscripts or through brief discussion (usually right before being summarily dismissed) by later thinkers such as Aristotle. Some of them held ideas that sound downright preposterous to us, such as that the fundamental nature of all things is water (according to Thales of Miletus, sixth century BCE), air (following Anaximenes, also sixth century BCE), or fire (this was the idea of Heraclitus of Ephesus, circa 500 BCE). For that matter, Pythagoras and his followers (sixth and fifth centuries BCE) may have thought that numbers are the fundamental constituent of the world—though it is debatable whether they meant that literally (as crazy as it sounds) or whether, in a more sophisticated and modern fashion, they maintained that mathematics is a crucial tool to uncover deep truths about the universe.

Then again, perhaps they weren't that crazy after all: modern physicists also keep looking for the ultimate foundations of matter, and think they have found it in subparticles called "quarks"—with bizarre names such as "up," "down," "beauty," and "charm"—or in even stranger things called "strings," stuff that vibrates at different frequencies, thereby determining the physical properties of quarks themselves.[3] Of course, modern physics is, we think, much closer to understanding the actual reality of matter than the pre-Socratics could have imagined, and modern physical theories such as quantum mechanics are based on sophisticated mathematics and innumerable experiments. Still, things had to start somewhere.

In fact, even closer to the truth than Thales and his colleagues was another group of pre-Socratics, the so-called atomists. These included Leucippus (also from Miletus, around 440 BCE) and Democritus (from Abdera, around 410 BCE), who thought that the world is made of indivisible fundamental units, which they called "atoms" (indeed, "atom" means "uncut"). The ancient atomists sometimes get credit (especially in pseudoscientific circles) for the wrong thing and do not get enough recognition for their true achievement, which they share with other pre-Socratics. Leucippus and Democritus did *not* anticipate modern physics, not in the sense that they thought of atoms in anything like the modern sense of fundamental particles. But even if one gives them credit for the idea that the world is made, at bottom, of one kind of stuff, the pre-Socratics had no concept of what sort of things atoms (or quarks, or strings, as we would say today) were and—more importantly—had no clue about how to study them. So to claim that the Greek atomists "anticipated" modern subnuclear physics is as preposterous as the fashionable trend of finding "parallels" between obscure pronouncements of Eastern mysticism and quantum mechanics.

However, both the atomists and other pre-Socratics were onto a few other things that turned out to be very important for the development of what we today call science. To begin with, they were interested in asking fundamental questions about the nature of the world. That is, unlike their Egyptian and Babylonian predecessors (as well as their Roman successors), the emphasis was on knowledge for knowledge's sake, regardless of the possible material benefits to humankind. I do not mean to oversimplify things and draw a sharp distinction between basic and applied science; the two are certainly interconnected and inform each other, with basic science providing the fuel for applications and new technological advances, applied science sometimes in turn making it possible to carry out better fundamental science (think of the invention of radars, which made possible the entire field of radioastronomy). Nonetheless, it is also important not to downplay the distinction: science (and philosophy) began when the pre-Socratics shifted their efforts from medicine and agriculture (not to mention religion and mysticism) to physics and astronomy because they were interested in how the world actually works.

The tension between basic and applied science is still felt by every practicing scientist today. In the United States, the National Science Foundation gives out grants for basic scientific research, while a host of other federal agencies, including the U.S. Department of Agriculture, the National Institutes of Health, and the Department of Energy, give money for more applied scientific projects. Needless to say, NSF's budget is a puny amount compared to any of the other agencies, and perhaps so it should be; after all, that budget is paid for by taxes levied on citizens, who are presumably more interested in cures for diseases than on figuring out, say, the mating rituals of a particular species of moths. The tension manifests itself more obviously when basic scientists have to give the National Science Foundation a justification of why they want the money they are asking for, in terms not just of the specifics of their research projects, but of the "broader impact" of what they are doing. In other words, non-applied scientists get their money if they can make a convincing case that what they do may, someday, actually have practical benefits. This often leads to (unintentionally) comical short paragraphs tacked at the end of an otherwise lengthy and detailed grant proposal, in which the scientist who wishes to study the above-mentioned sexual habits of moths waxes poetic about how this will somehow further our understanding of animal sex in general (hence, including humans), or of why invasive species are so effective at colonizing new habitats (if the moth happens to be an invasive species), or of how global warming affects the mating rituals of animals across the globe. I'm being a bit facetious here, but only a bit. The point is that what the ancient Greeks started in terms of conceptual separation between fundamental and applied science is still with us today, which is no small influence over the course of the intellectual history of the past 2,500 years.

The second reason the pre-Socratics made a fundamental contribution to science is their willingness to accept conclusions that go against commonsense. It is often said that the scientific method is nothing but "commonsense writ large." That may be true about some of science's methods, but the *results* of science are often far from commonsensical, and the most spectacular scientific discoveries positively defy our intuitive understanding of the world. Just think of the conclusion reached by modern subatomic physics that solid objects are made largely of empty space! The reason they feel solid to us is because the molecules of which they are made are bound together by strong forces at the submicroscopic level, forces that we can neither perceive nor penetrate. Nonetheless, the actual "stuff" that makes up the objects of our everyday world is pretty rarefied so that if we could observe, say, a table at the atomic level, we would see a few atoms here and there, separated by large regions in which there is no matter at all.[4]

Similarly, some pre-Socratic philosophers were willing to follow their reason wherever it led them, even at the cost of literally denying the evidence of their senses. True, in most cases this resulted in what with hindsight we consider rather preposterous positions. Perhaps the most famous one is the conclusion reached by Zeno (circa 450 BCE), a student of Parmenides. Parmenides already had the guts to teach that change is an illusion, on the ground that change entails the transformation from nonbeing to being, which in turns entails a logical contradiction. Since logic can't be wrong, then it is the apparent reality of change that must be rejected. Zeno went a step further and contested the ostensibly incontestable existence of motion. His reasoning was pretty airtight (unlike Parmenides'): Zeno said that in order to, say, go from his home to the agora (the central square in the city where all sorts of interactions, from commercial transactions to philosophical discussions, usually took place) he would have to first cover half that distance. But to do so he would have to cover half of *that* distance. And so on, ad infinitum. Since it is impossible to physically go through an infinite sequence of steps, then it must be impossible to get to the agora (or to any other place, mind you).

Now, our first reaction in hearing this is that Zeno must have had more than one loose screw in his brain. But by all accounts Zeno was a perfectly functional human being, which means that he knew very well that he could go to the agora, and presumably he did so all his life. What gives, then? The point is that Zeno actually put his philosophical finger on a real problem: how do we make sense of infinite series? It took Newton's and Leibniz's independent codiscovery of infinitesimal calculus in the early eighteenth century to solve Zeno's paradox, and by all measures of intellectual achievements, having posed a problem that had to wait twenty-one centuries to be solved certainly is an impressive feat!

There is a third reason the pre-Socratics distinguish themselves so sharply from their own predecessors and even many of their successors until modern times: they rejected supernatural explanations for the way the world works. For the atomists, for instance, both the physical and the biological worlds were the result of atoms moving in a void, without organizing intelligent principles. The Milesians (i.e., Thales, Anaximander, and Anaximenes) were also what today we would call materialists, not in the vernacular sense that they were interested only in material gain, but in the philosophical sense that they thought that matter is all there is. While scientists as individuals are certainly not committed to philosophical materialism (i.e., one doesn't have to be an atheist to be a scientist), they all in fact use what in the last chapter we called "methodological naturalism": science is based on the assumption that there are natural, material causes for the phenomena we study, and therefore that all scientific explanations are inherently naturalistic. We owe this crucial insight, which represents one of the foundations of modern science, to the pre-Socratics. As we have seen, this pre-Socratic move is still found to be highly objectionable by defenders of intelligent design creationism—not surprisingly, considering that they would like to see supernatural explanations reintroduced in science. They do not seem to realize that we would simply not be doing science anymore.

The last great contribution of the pre-Socratics to the foundations of science is that they realized that they couldn't just go around *telling* people what reality is like. They had to defend their conclusions using logic and rationality. Although they didn't engage in experimental scientific work (as we shall see in a moment, we have to wait until Aristotle for people to gather systematic observational data on which to base their conclusions), the pre-Socratics embraced the radical notion that one needs to defend one's ideas against the competition. The use of reason in this sense has been a hallmark of both Western philosophy and science ever since, and it sharply distinguishes them from other forms of thought, such as Eastern mysticism and much of religion.

The pre-Socratics' influence extended most immediately and directly to the generation of philosophers that operated in Athens around the turn of the fourth century BCE. One of them made immense contributions to philosophy, but refused to engage in science; a second initiated the split between philosophy and science that still exists today; and the third one was both a highly influential philosopher and the first scientist in anything resembling our sense of the term.

Socrates, Plato, and Aristotle: Establishing Philosophy as a Dominant Form of Thought

If philosophers ever played with fantasy leagues on the model of sports fans, by far the most voted trio would have to be the one constituted by Socrates, Plato, and Aristotle. They lived and operated largely or exclusively in Athens around the turn of the fourth century BCE, and they were each other's pupils: Aristotle (384–322 BCE) studied with Plato before starting his own *Lyceum* school; Plato (424/3–348/7 BCE) in turn had been Socrates' student before the latter's death in 399 BCE, some time after which Plato started his *Academy*; as for Socrates himself (470–399 BCE), we have already seen that his contributions to philosophy are so fundamental that everyone who came before him is simply referred to as "pre-Socratic."[5]

With respect to their contributions to the origin of science, however, these three philosophers could not have been more different. Socrates never actually engaged in the natural sciences, devoting himself instead to ethics and moral theory. Indeed, he started a conceptual separation between matters of fact and issues of values that philosophers have begun to question only during the latter half of the twentieth century. But that's another story.[6] Plato did have something to say about science, but much of it amounted to setting up problems that later philosophers had to overcome. Let me start by first acknowledging that Plato elaborated a cosmology (in his *Timaeus* dialogue) that was influential through the Middle Ages. He had an account of creation (though scientifically problematic, as we'll see in a moment) and thought the earth was spherical. He arrived at a relatively sophisticated model of planetary motion (of course featuring a stationary earth at the center of things) and even commented on human physiology and medicine. It is often said that Plato discounted the value of observation, but this is actually not the case, though it will be Aristotle who will make empirical data an integral part of his natural philosophy.

Plato even attempted a synthesis of the pre-Socratics' view of the fundamental elements (air, fire, water, and earth) with those of the mathematical mystics of the Pythagorean school. The Greeks had already figured out that there are only five regular solid geometrical figures possible (all others are combinations of the fundamental five), and Plato associated each one of them with one of the fundamental elements, taking seriously the idea that mathematics can provide a key to the understanding of the nature of reality. This idea is still at the basis of much scientific practice. Of course, you may have noticed a problem with Plato's scheme: he had four elements and *five* geometric solids to match with each other. Undaunted, he simply decided that the most complex geometric figure, the dodecahedron, represents the entirety of the cosmos. Neat, albeit with no empirical foundations whatsoever.

Indeed, a clear conceptual priority of logic over observation did mark Plato's thought and reverberated all the way into the eighteenth century, with the philosophical debate between rationalists like René Descartes and empiricists such as David Hume. The details of that debate are beyond the scope of this book, but an understanding of the basics is very helpful in order to gain an insight into the nature of science and its gradual separation from philosophy.[7] In essence, the rationalist position, which can be traced back to Plato (though the term "rationalist" will identify a much later philosophical school), while not denying that observations are important, emphasizes the incontestable fact that our senses can be mistaken. Just think of optical illusions. For Plato, therefore, true knowledge can be obtained only by the use of reason, even though sometimes our rational conclusions seem to fly in the face of what our eyes are telling us.

The empiricists, chiefly represented by Enlightenment figures such as John Locke, George Berkeley,[8] and the above-mentioned Hume, argued instead that ultimately all knowledge must come from the senses, which inform and gradually shape how our minds conceive the world. Locke's famous image of the mind at birth was that of a tabula rasa, an empty slate over which experience begins to write.[9] A more sophisticated empiricist, Hume came up with the infamous "problem of induction" (though he didn't use this term), which is actually what got philosopher Karl Popper to think about the demarcation problem (i.e., the difference between science and nonscience) that we considered in the introduction. Hume reckoned that science makes use of the two fundamental types of logical reasoning that have been known since Aristotle: deduction and induction. Deduction allows one to arrive at guaranteed conclusions *if* the premises are true and the structure of the argument is sound. For instance, take the standard (Aristotelian) example of a syllogism, a type of simple deductive reasoning:

Premise 1: All men are mortal.
Premise 2: Socrates is a man.
Conclusion: Socrates is mortal.

If we have good reasons to believe the first two premises, the conclusion is inescapable. And of course Socrates did die, by order of the Athenian government, in 399 BCE.

But, says Hume, how do we *know* that the two premises are, in fact, true? Socrates could be a woman or an alien from outer space (Hume did not use these specific examples); and it is at least theoretically possible that not all men are mortal (this is probably biologically impossible, but the point is that immortality does not violate any rule of logic). Hume's answer was that we can know these things only through direct observation (sense data) or induction, the process of generalizing from a number of observations (more sense data) to broader instances. We think we know that all men are mortal because we've never encountered an immortal one. But it is conceivable that we might do so tomorrow (though I, for one, wouldn't know how to tell an immortal from a mortal). And we know that Socrates was a man because he looked and behaved like a man (though he may have been a cyborg from the future). The idea is that induction is always fallible, and its conclusions tend to be probabilistic. Deduction yields foolproof conclusions, but at some point or other must rely on induction. This "problem of induction" is serious enough to have caused philosophers plenty of nightmares, but Hume's position that even deduction must at some point be based on gathering (possibly faulty) sense data was particularly problematic for rationalists like Descartes, who wanted to achieve certain Knowledge (with a capital *K*). Instead, the best anyone can do is to arrive at provisional conclusions. This is why scientific theories are always tentative: because they are based on a combination of deduction and induction (and a few other ingredients), and the inductive component is always open to revision.

Now, modern philosophy has gotten a couple of good handles on the problem of induction and the apparent conflict between rationalism and empiricism. The first such handle was provided by one of the most influential philosophers of all time: Immanuel Kant. Kant started out his career in Konigsberg (Prussia) as a run-of-the-mill rationalist, writing decent philosophical papers, which, however, weren't terribly exciting.[10] Then, in his own words, Kant was "roused from a dogmatic slumber" by reading Hume, and the effect was to give "quite a new direction to [his] researches in the field of speculative philosophy."[11] Kant's idea was that neither rationalism nor empiricism got it entirely right, though elements of both were to be retained. What was missing was that human experience of the world (and hence, we can add, our knowledge of it) depends on *three* elements, not two: there are the sense data (emphasized by the empiricists); and there is rational argumentation (championed by the rationalists); but there is also the fact that we experience the world through the use of some conceptual filters—Kant calls them "Ideas"—that we derive neither from experience nor from logic. These filters include notions such as space and time. We don't learn about space, say, from experience, since we already need a grasp of the concept of space to *make sense* of any sensorial experience. But a fortiori we don't arrive at the idea of space from a logical analysis of how the world works either. Kant's is a beautiful idea, arguably one of the most brilliant intellectual insights in human history and one that still provides the background for our discussions on the nature of knowledge.

But if the Ideas are derived neither from experience nor from reason, where do they come from? Kant couldn't say, but a modern biologist would argue that our intuitive notions of some aspects of the world are the result of our evolution as sentient beings. We know that other primates, for instance, also behave as if they had an innate grasp of space, and it makes sense that such instinct would evolve by natural selection to help the animal navigate its world effectively, but I will not pursue this line of inquiry here, because it would lead us very far afield of our current interest.

Back to Plato, then. We have seen that he realized that both rational thinking and observation of the actual world are important in acquiring knowledge, but that he gave precedence to the first over the second. This means, then, precedence of deductive over inductive reasoning, for while thinking and observing are sources of knowledge, and deduction and induction are forms of reasoning, deduction is given primacy by the rationalists (and by Plato) and induction by the empiricists. Interestingly, it is this Platonic choice that generated one of the most enduring—and questionable—ideas of Western philosophy: that of an ideal world that is somehow more real than, well, the "real" one. Plato introduced his so-called theory of forms with a beautiful analogy in his dialogue *The Republic* (book 7), in which he imagined a cave where people are chained facing a wall. They are born in the cave and are never allowed to leave. They can see something of the outside world because of the shadows projected from the light at the entrance of the cave. It is important to understand that for the people in the cave these shades *are* the reality. They are not aware that they are just a pale reflection of a much more complex (and "real") world.

Now, says Plato, imagine that one of these men (a philosopher, predictably) somehow frees himself of the chains and ventures outside. His senses will be overwhelmed with colors and sounds and smells, and at first he will have a hard time understanding what is going on. However, eventually he will absorb the situation and will be eager to go back to the cave to tell the others about his discovery. But, of course, they will not believe him—they will mock him, even. They simply cannot conceive of this truer reality that he claims to have accessed and will likely dismiss him as a pompous intellectual who thinks he knows more than they do. The joke, of course, is on those who will refuse the true knowledge brought in by philosophy, the only way, according to Plato, to get outside of the cave of our senses and directly access the world as it really is.[12]

Through his parable of the cave Plato wants to convince us that the reality that we experience is but a pale reflection of a higher, more perfect reality, which he calls the world of ideas.[13] Where did he get this strange notion? Remember Plato's preference for deduction over induction: if there is a perfect world of ideas out there, we can derive (deduce) perfect knowledge about it, thereby realizing the rationalist goal. We can also make sense of the actual world in terms of the ideal one. For instance, the various dogs we observe are but imperfect instantiations of the ideal dog; the same goes for the ideal chair versus actual chairs, the ideal philosopher versus actual philosophers, and so on (Plato did not address the question of whether, say, there is such a thing as an ideal disease, or ideal garbage). In other words, Plato thought that the abstractions came first, that they had what philosophers call ontological priority (i.e., they exist prior to their individual manifestations). Most of us would find the reverse obvious: arrive at the general idea of dog, chair, philosopher (disease, garbage) by abstraction from the observation of particular instances (as Plato's pupil, Aristotle, later did). If one proceeds by generalization (which is a form of induction), as Plato refused to do, it becomes clear that the individual objects are the real things and that abstract ideas are just that—abstract. There is no cave, and philosophy doesn't lead us out of it after all.

There is one more move that Plato made that represented a step back in the context of the development of science, a step that took until the sixteenth century to begin to correct (ironically, by the rationalist Descartes): he reintroduced supernaturalism to our investigations into the ultimate basis of reality. As we have seen, the gods were expelled from philosophical inquiry by the pre-Socratics and in particular by the atomists. The goal was to explain nature on nature's own terms, an objective that characterizes modern science as well.[14] But Plato would have none of it: his cosmogony (the explanation of how the world came to be) included the action of a demiurge, a minor god who made the world in the best way he could out of the available materials. Although Plato didn't bother to tell us where these materials came from, at least his adoption of a designer who is intelligent yet limited in what he can do was actually more sophisticated than most successive (Christian-informed) conceptions of intelligent design, because the actions of Plato's demiurge were compatible with the existence of an (obviously) imperfect universe. As Leibniz would put it much later, we live in the best of all *possible* worlds, which is very different from saying that we live in a perfect world.

Some of the problems with Plato's philosophy began to be corrected by Aristotle, who can with good reason be considered the first scientist in any sense even remotely approaching our conception of science. Aristotle was by all accounts a remarkable figure, interested in what today encompasses the whole of science and philosophy. He invented the concept of metaphysics[15] and wrote about moral theory in a manner that still today provides the obligatory beginning of any serious discussion on ethics. He also wrote about a stunning variety of scientific fields, including cosmology, biology, physics, and meteorology. He defended the idea that the earth is a sphere and reported calculations that estimated it to be about 45,000 miles in circumference, not very close to but in the same ballpark as today's figure of 24,902 miles. Aristotle produced a theory of motion and one of the nature of space as well as an account of the structure of the cosmos, and he was fascinated by the functioning of the eye, producing a theory of light and vision that was still debated in the seventeenth century. To disregard Aristotle's contribution on the ground that most of his theories turned out to be wrong and were eventually superseded is to seriously misunderstand the nature of history and of intellectual progress: without the benchmarks set by Aristotle we simply could not have built the much more sophisticated science of today.

Despite being a disciple of Plato (or perhaps because of it), Aristotle rejected Plato's notion of a realm of ideas of which the reality that we experience is but a pale reflection. In a move that essentially reestablished the importance of observation and inductive thinking so crucial to science, Aristotle correctly reversed Plato's line of thought, suggesting that we can think of, say, an "ideal" dog only as an extrapolation of the mind based on the observation of a number of real dogs. But for Aristotle real dogs are all there is, and properties (such as "dogness") cannot be abstracted from the objects with which they are associated. This is, of course, not to say that Aristotle did not value deductive logic—after all, he is considered the founder of the entire discipline of formal logic, which is based on the idea of deduction. But it is fair to say that in this and many other respects Aristotle had a view of knowledge that is much more in synchrony with our modern understanding of science than Plato's.

For instance, Aristotle understood that biology, in particular, is a complex combination of observations and explanations: without intense observation of the biological world one simply cannot begin to shape informed ideas about the causes of biological phenomena; accordingly, he spent years on the island of Lesbos[16] observing and cataloging marine life, and his *History of Animals* lists over five hundred species, the beginning of the modern science of taxonomy. This is an approach not much different from that of Charles Darwin, who embarked on his famous voyage on the *Beagle* as a naturalist precisely to have the opportunity to make unprecedented observations of biological organisms in their natural settings; that voyage eventually resulted in his theory of evolution by natural selection.[17] But for Aristotle, as for Darwin and modern scientists, science cannot be limited to natural history: it has to include causal explanations of the phenomena of interest. And here is where Aristotle's thinking took a highly original turn, one that—for good and for bad—has influenced the development of science for more than two millennia and that is still relevant to modern discussions of the relationship between science and religion.

Aristotle developed a sophisticated conceptual framework of causality, a framework that informed every question he posed and how he went about answering it. For Aristotle there are four, and only four, types of causes, and he called them material, formal, efficient, and final. The classic example is the one of a statue: the broad question of what "causes" the statue to come into existence is broken down by Aristotle into four questions and corresponding answers. The *material* cause refers to what the statue is made of (say, bronze); the *formal* cause is the specific shape of the statue (say, that of a philosopher); the *efficient* cause has to do with how the statue comes into being (through the work of a particular sculptor); and the *final* cause accounts for why the statue is being made to begin with (perhaps because the city wishes to honor a great thinker). (The example is more convincing if one uses a sports star or a movie celebrity, but you get the gist.)

There are at least two important consequences of this way of framing the issue of causes, one of which got Aristotle in perpetual trouble during the Middle Ages, while the other led to a problem for science that was resolved only during the nineteenth century. The first consequence is that Aristotle broke with Plato on the issue of the soul: Aristotle identified the soul with the form of an organism (inorganic things don't have souls, plants have only "nutritive" souls, animals are characterized by both nutritive and "sensitive" souls, and humans sport the addition of a "rational" soul). Since for Aristotle properties cannot exist independently of objects, it follows that souls die with the body. This, as one might imagine, did not go over well with Medieval Scholastic philosophy, with many Christian philosophers—after the rediscovery of Aristotle's work—having to reconcile their love for the Greek thinker with the mortal blow he dealt to one of the fundamental concepts of Christian doctrine. The dualism between body and soul (or, we would say today, mind) that Aristotle rejected persisted in Western thought from Plato through the Middle Ages, and it has not entirely disappeared yet. It was only Descartes' failure to account for how an immaterial soul could possibly be connected to and influence a physical body that finally began to crack the edifice, leading to the modern scientific conception that all mental activities are ultimately physical.[18]

The second issue has to do with the concept of final cause, which was crucial for Aristotle in that it addresses *why* things are, as opposed to merely explaining *how* they come into being. For Aristotle *why* preceded *how* in the same way in which no matter what sculptor, material, and shape may result in the statue of a philosopher, none of it is going to even get started unless the city makes the decision that it *wants* said statue. Aristotle applied this reasoning to everything, not just to humanmade artifacts, so that for him it made perfect sense to ask what is the final cause (i.e., the reason) for the existence of living organisms or even of the universe itself. This line of argumentation quickly leads to thinking in terms of (supernatural) intelligent design for both living beings and the universe, a way of thinking that is now soundly rejected by modern science. (Aristotle was not, however, a supporter of intelligent design in anything like the modern sense of the term: he thought of the beginning of the universe in terms of an uncaused "first cause" that was identified with God only by later Christian thinkers.)

Indeed, Francis Bacon later argued that science, in order to separate itself once and for all from philosophy, had to face the fact that talking about final causes is vacuous; still today, many scientists will claim that science deals with *how* but not with *why*, and that the latter is best left to the philosophers (usually this comment is accompanied by an imperceptible smirk, to make sure the interlocutor understands that science deals with the important stuff and that philosophy is for people who can afford to indulge in intellectual games). We only had to wait two and a half centuries after Bacon, however, for final causes to reenter science in a triumphal and unexpected (and wholly non-Aristotelian) way: Charles Darwin's work showed that there is an important sense in which one can ask *why* there are eyes, say, and not just *how* they are built. There are eyes because, under certain environmental conditions, it is advantageous for animals to capture light in order to determine basic features of their environment, like where possible prey may be hiding or predators waiting. In other words, the new theory of natural selection allows biologists to again pursue all of the types of causal explanation envisioned by Aristotle.[19]

What about Math?

Before we leave the ancient Greeks, we need to briefly consider another fundamental contribution they made to the foundations of modern science: mathematics. I have already mentioned Pythagoras's mathematical mysticism, which influenced Plato's theory of forms and still resonates today in (rather grandiose, if philosophically dubious) assertions by theoretical physicists like Stephen Hawking that their research provides a way to "look into the mind of God." Aristotle, ever seeking to strike a moderate position, thought it more likely that mathematics were a useful tool with which to approach certain problems, but no more than a tool that can be integrated with or replaced by others depending on the nature of the specific inquiry.

Nonetheless, most ancient philosophers were fascinated by the fact that one could use deductive logic to prove mathematical and geometrical theorems, especially after the elegant work of Euclid, which still provides the foundations for plane geometry. This eminently Greek idea that deduction is superior to induction and that the best science ought to be based on the former survived, as I mentioned, until the scientific revolution of the seventeenth century, and traces of it can still be observed today. Regardless, even by modern standards the ancient Greeks' penchant for applying mathematics to the study of nature delivered significant results. Examples abound, and a good discussion of the Greeks' accomplishments in math and geometry can be found in David Lindberg's book *The Beginnings of Western Science*. A few instances will suffice to get an idea of how far things had progressed by the onset of the Middle Ages. Eudoxus of Cnidus (fourth century BCE) managed to create the first known geometrical model of planetary motions, which accounted for the so-called retrograde movements of planets like Mars, during which their apparent progression in the sky at times reverses direction, only to resume the usual behavior a few days later. Plato's successor at the Academy, Heraclides of Pontus, figured out that the earth rotates on its axis every twenty-four hours or so, and a few decades later Aristarchus of Samos came up with the idea that it is the sun, not the earth, that is situated at the center of the solar system—anticipating Copernicus by 1,800 years. Aristarchus had also arrived at a method to estimate the ratio between the distances of the sun and the moon from the earth, although the results were far from accurate. After Aristotle's early attempt, Eratosthenes (circa 235 BCE) calculated the circumference of the earth and obtained a value that was within 20 percent of the currently accepted one.

Also in astronomy, of course, Ptolemy (no relation to the Egyptian dynasty to which Cleopatra eventually belonged) was the metaphorical king. His system of epicycles accounted so well for the movements of the planets that it was not seriously questioned until Copernicus, and it took Galileo and Kepler to definitively show that it was wrong. Both Euclid (circa 300 BCE) and Archimedes (287–212 BCE) were interested in the physics of solids, and the latter in particular worked out an elegant proof of the law of levers. In optics, several ancient philosophers attempted to come up with a theory of vision, ingenious ones are attributed again to Euclid and to the very same Ptolemy. Once more, we had to wait until the scientific revolution and Newton's work to get significant further improvements.

There continue to be, and perhaps will always be, areas of science where mathematical approaches are not terribly useful and where in fact an obsession with quantification may be positively deleterious. For instance, sciences that investigate highly complex phenomena (psychology, sociology) or that deal with mostly historical subject matters (paleontology) are not amenable to the sort of powerful but relatively straightforward analytical techniques that enthralled Pythagoras and Archimedes. Even where statistics (by its own nature a very different kind of mathematical approach) is useful to these sciences, it is as an auxiliary tool to attempt to extract patterns from messy data sets, and may hinder the study of individual (as opposed to population) differences. Galileo famously said that the book of nature is written in mathematical language, but this is a metaphor whose actual meaning or value remains hotly disputed. "Nature" is, of course, not a book, and it isn't written in any language whatsoever. Nevertheless, it is indisputable that we simply would not have modern science without mathematics and that the ancient Greeks from Pythagoras to Archimedes were instrumental in gifting humanity with this insight.

Intermezzo: Romans, Medievals, and Arabs

Having considered several fundamental contributions made by ancient Greek philosophy to the beginnings of science, we need to see how the latter finally severed its umbilical cord from natural philosophy, a transition that we will examine in the next chapter, chiefly through the work of a few scientist-philosophers whose lives spanned the three centuries of the so-called scientific revolution. Before that, however, we need to (very briefly) take a look at what happened to science between ancient Greece and the post-Renaissance revolution.

The Romans inherited much knowledge, practical as well as theoretical, from the Greeks, who had also managed little-known but nonetheless impressive discoveries in human physiology and anatomy, and hence in medical practice. The Greco-Roman world was already affected by and conscious of the very same "demarcation problem" between sound science and flimflammery that interests us in this book (although certainly they would not have put it in those terms). The famous Hippocratic oath, named after Hippocrates of Cos (460–370 BCE), was essentially an attempt at regulating medical practice and limiting the doings of charlatans. Authors writing in the Hippocratic tradition made the same move that we saw the pre-Socratics make: gods may exist, but they do not account for disease or how to cure it. In fact, medical theory and practice had quickly advanced to a level that we normally associate only with the Renaissance and the birth of modern European universities: the human body was being dissected already in the third century BCE, and Erasistratus (circa 304 BCE) had written out a complete theory of how digestion works. Herophilus (circa 260–250 BCE) figured out both the connections between the brain and the spinal chord and the distinction between veins and arteries.

This chapter is not meant to be a chronological history of scientific discoveries, as my emphasis is on the various steps that led to the conceptual foundations of what we call science. Yet we cannot skip the contribution of Galen (129–210 CE), who was a physician to various Roman emperors. While his summary of specific medical knowledge covers several volumes in its modern edition, his factual conclusions and theories have, of course, been superseded. But he was among the first to assert yet another fundamental principle of modern science: reductionism. Galen concluded that the best way to diagnose a disease was to look not at the entire body, but to focus instead on whatever organ appeared to be affected—that is, he reduced disease to the specific malfunction of individual body organs. Galen lived in a time in which human bodies could no longer be dissected for research because of religious proscription; undaunted, he introduced the revolutionary idea that we can use animal models instead, a practice that is still today very much at the center of medical and pharmaceutical science (and, of course, of ethical disputes).

With the remarkable exception of Galen (who was not actually Roman, though he spent the last part of his life in the imperial capital), there is little to be said about the conceptual history of science from the Roman period through the Middle Ages. This is not because learning was not valued or because of the absence of great scholars, but rather because of a peculiar combination of political and economic factors as well as a particular attitude toward scientific inquiry that characterized the Romans. David Lindberg cites a famous quip by the Roman playwright Ennius (239–169 BCE) to the effect that his compatriots were inclined "to study philosophy, but in moderation."[20] This summarizes the Romans' outlook wonderfully. While they could sport a number of philosophers and commentators (Cicero, Seneca), natural historians (Pliny the Elder), and even what today we would call science popularizers (Lucretius), for the Romans intellectual pursuit was something one does for diversion and amusement when one is not absorbed by the really important things: politics and war. To some extent, of course, this is true in general: from ancient Athens to modern times, intellectual pursuits could be indulged in only when there is peace and economic prosperity, and as soon as one or the other began to falter, resources were quickly shifted away from science (not to mention philosophy, which even in modern academia isn't exactly awash with dollars). But the Romans' entire economy was predicated on an almost steady wave of conquests, which meant that war was never far away. In turn, this translated into seeing intellectual inquiry as little more than a relaxing hobby, even at the height of the economic and military power of the empire (the same, of course, cannot be said of applied sciences, particularly engineering, in which the Romans excelled like few others in the entire history of humanity; then again, one hardly needs to point out the practical benefits of roads, bridges, and aqueducts for the Roman military machine).

The situation was different, although science and philosophy did not fare any better, during the Middle Ages. For many years now there has been controversy among scholars on whether the classic characterization of that period as the "Dark Ages" is in fact appropriate, given that some interesting scientific discoveries were made during its span (though mostly toward the end, in a prelude to the Renaissance). Be that as it may, it is still hard to conceptualize the Middle Ages as an enlightened period for humanity in general and for intellectual pursuits in particular. For one thing, economic prosperity was a thing of the past for most people in Europe after the fall of the Roman Empire, though of course there were exceptions, limited to particular locales and short time spans. More importantly, the Christian Church took an attitude that was no more helpful to science and philosophy than the Romans' had been, and, in fact, was arguably significantly worse.

The tone was set by Augustine (354–430 CE), who argued that philosophy (including natural philosophy, i.e., science) was to be encouraged only as a handmaiden to religion. After the rediscovery of their works, it was fine to study Plato and Aristotle (not so much the more radical Epicurus), but one needed to keep in mind that when their philosophizing contradicted the doctrine of the church, it was the latter that prevailed—by iron fist, if not by logic, as the burning at the stakes of Giordano Bruno (as late as 17 February 1600) reminded everyone, particularly a certain Tuscan astronomer named Galileo Galilei, who a few years later escaped the same fate only because he happened to be on good terms with the pope (having friends in high positions has rarely hurt scientists, or anyone else for that matter). Given this climate, most of what we have in Europe during the Middle Ages is a tradition of copying ancient manuscripts (resulting in what are sometimes stunning works of art from a purely aesthetic viewpoint, regardless of the contents that were being copied) and much philosophizing in the so-called Scholastic tradition. Scholasticism has a bad reputation in philosophy, its practitioners being the very same people who often engaged in discussions about extremely fine points of theology and against whom Francis Bacon eventually directed his invectives. As far as we are concerned here, Scholasticism is irrelevant because it contributed nothing worth mentioning to the conceptual history of science, despite having attracted some remarkable minds of the Western tradition, such as Thomas Aquinas (1225–74).[21]

Before adjourning, we need to briefly consider the contributions of yet another tradition that intersected the Middle Ages and altered the history of Europe, both culturally and militarily: Islam. As in the case of the debate about the "Dark Ages," much controversy has marked the recent reevaluation of Islam's contributions to science—particularly in light of the ongoing cultural and military clashes between the West and fundamentalist Islamic groups. The fact of the matter is, however, that once again we see little in the way of either conceptual advances or even genuine discoveries and much in the way of copying and translating other people's work.[22] The translation in Arabic of Greek works occurred because the decaying Eastern Roman Empire (with its capital in Byzantium) came into contact with Syria, which in turn was quickly annexed by the aggressively expanding Islam by 661 CE. This cultural contact—violent as it was—turned out to be of fundamental importance for the history of philosophy and science because without it a great number of manuscripts would have been lost forever. The West returned the favor to humanity a few centuries later, when Islam's own decline between 1085 (the fall of Toledo) and 1258 (when the Mongols took Baghdad) and beyond led to yet another transfusion of knowledge—this time one that fueled the cultural recovery of Europe and culminated with the invention of universities first and the development of the Renaissance later.

What is most relevant to our discussion, however, is that in Islam science was again seen as handmaiden to religion, a concept probably arrived at independently from the Christian analog, but equally pernicious to free inquiry. Indeed, Muslims have been suspicious from the beginning of all foreign learning (often labeled "rational," not to be read as a term of endearment) and with few exceptions have always considered intellectual activities such as philosophy and science useless or, worse, dangerous to the faith (not entirely an irrational position, when considered in itself). Predictably, public understanding of science in Islam suffers still today. Even semisecular Turkey is infested by fundamentalist creationists who find themselves aggressively defending the same inane positions as their American counterparts, and the situation is simply abysmal across the theocracies of the Middle East. But it is time now to take a breather, jump a few centuries into the future, and turn our attention to the beginnings of the scientific revolution.

CHAPTER NINE

From Natural Philosophy to Modern Science

The intention of the Holy Ghost is to teach us how one goes to heaven, not how heaven goes.
—Galileo Galilei, quoting Cardinal Baronius

Science is the great antidote to the poison of enthusiasm and superstition.—Adam Smith

Revolutions may take place overnight, but they usually require some preparation. The same is true for the so-called scientific revolution of the seventeenth century, which marks the official beginning of modern science.[1] What exactly happened during the late Middle Ages that allowed society to transition out of a period of European history where violence and poverty reigned supreme and where learning was largely the privilege of monks who focused almost exclusively on religious matters? As always with complex historical developments it is hard to untangle the causal network, but a couple of factors were clearly pivotal: war and urban expansion, and the two are themselves intricately connected.

For centuries after the fall of the Roman Empire, much of Europe was on the receiving end of armed invasions, with all the destruction and misery that such events routinely bring. But soon after the first millennium the tide began to change. First, Europeans fought against Islam's encroachment into their territories, and then they began to assault it in its own land, a sequence of events whose consequences still reverberate today, pregnant with social and political repercussions. An often bizarre collection of

European kings and princes mounted a total of nine "crusades" to gain control of the so-called Holy Land between 1095 and 1272. While ultimately unsuccessful, the crusades marked a turnaround in terms of political will and military initiative that affected the course of history from then on.

The other face of Europe's recovery soon materialized in the rebuilding and expansion of the cities. The Renaissance (from the Italian for "rebirth") that began in the fourteenth century, not to mention the scientific revolution that got started in the seventeenth, could simply not have happened without a revitalization of cities throughout the continent, from Bologna and Florence to Paris and Montpellier to Oxford and Cambridge. Not at all coincidentally, these are also the seats of the first modern universities, beginning with the one in Bologna, established around 1150. Ever since ancient Greece, learning has thrived in cities more than in rural areas, and the late Middle Ages and early Renaissance were no exception to the rule.

The rise of universities is, of course, a crucial event. Yet, a little appreciated fact about universities is that, in the beginning, they were not *places* of learning, but guilds of often itinerant teachers. These teachers could go to any of a number of cities known for their hospitality toward scholars and begin their trade, either independently or under the auspices of a school. As the number of teachers increased, there arose a need for them to become organized in order to gain and retain legal rights, including control over their intellectual property. When, shortly thereafter, teachers' guilds began to acquire properties, thereby establishing the actual physical places we now call universities, they also managed to secure a surprising amount of intellectual independence from local and even religious authorities. This was the embryonic version of what today is known as the principle of academic freedom. According to David Lindberg, even early on the courses of study offered at the university level could be taxing in terms of time commitment: after the Master of Arts degree (which at the time was no different from a doctorate), one would have to spend five or six more years to become a medical doctor, seven or eight years to become a lawyer, and a whopping eight to sixteen years to become a theologian![2]

Of course, a political and economic rebirth of Europe was necessary but not sufficient in itself to initiate the scientific revolution: genuine progress in learning in general and in natural philosophy in particular had to take place. This was no easy matter, largely—and perhaps ironically—because of the great influence that the Greek classics had on the intellectual climate of the early European universities. At the risk of oversimplifying things a bit, the greatest impetus to change was also the chief obstacle to overcome: Aristotle. Much of his work had been recovered in Europe by the early thirteenth century, largely thanks to the intermediacy of Arabic translations. The problem, in a nutshell, was this: on the one hand, Aristotle was an unavoidable benchmark in the intellectual development of the age, and his philosophy, once rediscovered, could simply not be ignored. On the other hand, a significant portion of what he wrote (for instance, his denial of the immortality of the soul or his conclusion that the world had always existed and therefore had not been created) were obviously at odds with church doctrine.

The problem soon erupted in a series of attempts by church leaders to curtail the influence of Aristotelian philosophy. In 1231, for example, Pope Gregory IX got personally involved with the issue of how the University of Paris (established only three decades earlier) should be regulated. He wrote that the faculty could not read Aristotle's books on natural philosophy unless they were first "examined and purged of all suspected error."[3] Lindberg justly notes that the surprising thing was that Pope Gregory admitted to the usefulness of *part* of Aristotle's writings, which were therefore not altogether banned. Still, the battle was on between academic freedom and religious censorship, which eventually claimed the life of Giordano Bruno and caused much personal trouble for Galileo.

An apparent solution to the problem was proposed shortly thereafter by Roger Bacon, who began teaching in Paris in the 1240s. Bacon was inspired, if that is the right word, by an old idea that can be traced to Augustine of Hippo (354–430 CE): "In Holy Scripture . . . all truth is rooted. I say, therefore, that one discipline is mistress of the others—namely, theology; for which the others are integral necessities and which cannot achieve its ends without them. And it lays claim to their virtues and subordinates them to its nod and command."[4] In other words, natural philosophy (i.e., science) could follow philosophy itself in a role as "handmaiden" of theology, a position that was of course in line with Pope Gregory's plan: take what can be useful from the new disciplines, but retain the religious authority's right to proscribe what sounds too inconvenient and potentially heretical.

Despite these rather unpromising beginnings, the intellectual floodgates were open, and a series of reinterpretations of the handmaiden doctrine eventually led to an effective independence of science from both philosophy and theology. Perhaps the first blow was dealt by Albert the Great (circa 1200–1280), the teacher of Thomas Aquinas. Albert's radical idea was that God works through the natural laws that He established to begin with. This neat, if theologically dubious, partial separation between the natural and the supernatural meant that science could focus on understanding the world based entirely on its own internal laws without encroaching on the territory of theology. Indeed, Albert went so far as to suggest that to talk of divine intervention in famous episodes of the Old Testament, like Noah's flood, would be an improper infringement on philosophy by theology. Aquinas, certainly one of the most influential theologians of all time, continued the same project and arrived at the conclusion that if philosophy and theology are at odds, that must be because of a human mistake (in *either* philosophy or theology!), not because of a true underlying conflict. This is essentially the doctrine reformulated by Pope John Paul II during the latter part of the twentieth century, when he explained (in his *Fides et Ratio*, Reason and Faith) why a scientific theory, if it is correct (in that case, the theory of evolution), can not be in contradiction with religion. Whether one buys into this sort of compromise or not, it historically began, and still continues today, the process of separation of religion and science that cannot but be beneficial to both.

Antiscience obscurantism did not simply bow to Albert and Aquinas and go away, though. When Albert the Great was still alive, in 1277, the bishop of Paris issued a condemnation of many philosophical positions, including several of those advocated by Aquinas. The latter's reputation suffered temporarily, but he was in fact made saint in 1323, and the 1277 articles were invalidated by another bishop of Paris in 1325. The fight went on, with ups and downs for the emerging discipline of natural philosophy, in part reflecting the appearance of strong personalities on all sides of the debate[5] and in part because of the fluctuations in the political and economic fortunes of various areas of Europe that served as the ideological battleground.

I do not wholly subscribe to what is sometimes scornfully referred to as "the great men theory of history," according to which major historical turns occur because particular individuals see much further than their contemporaries and push things forcefully in a given direction. Nonetheless, in the following section I will track the beginnings of the scientific revolution through the intellectual contributions of five scientist-philosophers: Francis Bacon (1561–1626), René Descartes (1596–1650), Galileo Galilei (1564–1642), Isaac Newton (1642–1727), and Charles Darwin (1809–82). These individuals were characterized, of course, by exceptional intellectual abilities that did, in fact, allow them to see further than others. But they also serve as reference points for the rapid development of ideas and discoveries that marked the sixteenth and seventeenth centuries, a development that catalyzed the final transition between philosophy and science.

Away from Philosophy, Forging Science

I will start my story with Francis Bacon, a man that has been described as cold, calculating, and corrupt, but also as a true Renaissance figure. He was highly praised by the likes of Voltaire, d'Alembert, and Diderot (the editors of the 1751 *Encyclopédie*), Leibniz, and Kant (who dedicated to him his *Critique of Pure Reason*, one of the most important philosophical works of all time). Bacon did, in fact, lose his high public office of lord chancellor under King James I (the very same who commissioned one of the most popular translations of the Bible) because of charges of corruption. The event, however, is considered by some to have been fortunate because it allowed (or, perhaps, forced) Bacon to concentrate his energy and time on writing instead of pursuing politics. His stature has become larger than life, with Mark Twain and Nietzsche defending the highly improbable hypothesis that Bacon was the true author of Shakespeare's plays.

As far as we are concerned here, Bacon is important because he was a classical transitional figure between the Middle Ages and the modern era. He studied in the then classic Scholastic manner, with its focus on the minutiae of Christian theology, and developed both an admiration for Aristotle and an aversion to the way Aristotelian philosophy had come to stultify intellectual discourse. He therefore set about to radically change the terms of the relationship among science, philosophy, and religion. He was done with the model of science as handmaiden to either philosophy or religion and wrote about how the new science was relevant to humankind mostly because, as he famously put it, "knowledge is power." Bacon was also the first modern writer to talk about the now familiar concept of technological progress, which ushered in the era of modernity to which science is inextricably connected. He was clearly regarded by his contemporaries as what we would today call "a public intellectual." As for Aristotle, Bacon was sufficiently brazen to entitle one of his major books *New Organon* in direct challenge to Aristotle's own *Organon* (the word means "tool").

For our goal of understanding the intellectual development of science, we need to discuss two fundamental aspects of Bacon's work: his famous "idols" of the mind and his emphasis on the use of the inductive method. To the first point, in the *New Organon*, Bacon tells us that before one can do serious science, one has to overcome a series of "idols" affecting the mind (the word derives from the Greek for "phantoms," i.e., sources of deception).[6] Bacon identifies four kinds of idols:

- **Of the Cave**, which derive from the vagaries of human culture (the particular "cave" in which we happen to be born and live), including all the prejudices that are associated with living in a particular time and place in human history. For instance, gender or racial biases, or allegiance to authorities, especially political or religious.
- **Of the Market Place**, meaning the ways people communicate with each other. Language can set its own traps against clear thinking, for example when we coin words for things that do not exist (like Aristotle's celestial spheres) and then proceed as if we had in fact identified real objects. (This problem is so common that it has acquired its own term in philosophy: "reification," from the Latin *res*, which means thing or object. To reify means to talk as if a certain thing were real, while in fact we simply coined a term that may not have a referent at all.)
- **Of the Theater**, by which Bacon meant (false) grand views of the universe that are bound to hamper genuine progress and understanding. Modern creationism would qualify as an idol of the theater.
- **Of the Tribe**, meaning the human tribe at large. These are the sorts of biases built by nature into human beings, like our overreliance on faulty sensorial experience (think of hallucinations, for instance), as well as more subtle and pernicious ones, like wishful thinking—the tendency to believe in certain things just because we would like them to be true.

The identification of these idols is a prime example of what Bacon called the *pars destruens* (literally, the destructive part). Bacon, however, believed that skepticism doesn't do much good if it is only negative; it is important to criticize and expose what is wrong, but then one also needs to put forth positive solutions. This *pars construens* (the constructive part) is presented in book 2 of the *New Organon*, in which Bacon talks about the method of induction.

We already encountered induction when we talked about how Aristotle thought that induction and deduction ought to be used in concert to advance our knowledge of nature. However, for Aristotle, as for the ancient Greeks in general, deduction (in which one arrives at a particular conclusion starting from broad premises) was the superior method, the only one by which we can achieve true and certain knowledge. In the *New Organon*, Bacon turns Aristotle's reasoning on its head, pointing out that the Achilles' heel of deduction is that if the premises turn out to be incorrect, the whole edifice crumbles in an instant. Induction, on the other hand, gradually builds from local observations to ever broader generalizations, providing—in Bacon's view—a much more solid foundation for scientific knowledge.

The import of Bacon's *pars construens* is not that he was right about how science proceeds or should proceed. Indeed, no practicing scientist today would recognize Bacon's method of slow and painstaking accumulation of facts as the most important part of what she does in the laboratory throughout her career. And modern philosophy of science has arrived at the conclusion that science works by a complex mixture of induction and deduction, among other ingredients (not the least of which is the expert intuition of which professional scientists are capable thanks to their many years of training). The point is that Bacon openly challenged Aristotle and that his influence reverberated for centuries. Charles Darwin, as we shall see below, considered his work to be driven by Baconian principles and defended it against his critics as a powerful example of the scientific method. More broadly, Bacon started a tradition of discussion about how science works, and that discussion evolved into what is now known as philosophy of science. However, it also marked the beginning of the divergence between philosophy and science that eventually resulted in a fully mature scientific enterprise. The scientific revolution had begun!

When Bacon died, in 1626, René Descartes was just thirty years old, with much of his work still ahead of him. Nonetheless, Descartes is in some sense an even more transitional figure between the Middle Ages and the modern era than Bacon himself. Descartes is chiefly remembered today as a philosopher, and yet he engaged in his metaphysical studies with the precise aim of making his physics acceptable to the Scholastic philosophers who at the time dominated the universities. Metaphysics is the branch of philosophy that investigates the nature of reality and therefore also the foundations of science. Descartes criticized his contemporary, Galileo, for having skipped metaphysics altogether, focusing too quickly on physics—that is, for daring to do science without philosophy. In a letter to Mersenne in 1638 he wrote: "Without having considered the first causes of nature, [Galileo] has merely looked for the explanations of a few particular effects, and he has thereby built without foundations." This is one of the last times in history that a scientist got seriously criticized for not doing proper philosophy. At the same time, Descartes was well aware of the risks of doing science a la Galileo, without justifying to the church why one's science was not in contradiction with accepted theology: Galileo had been condemned by the Roman Catholic Church just five years earlier, which in turn had prompted Descartes to dispense with the publication of his *Treatise on the World*, on which he had worked for four years, because it was likely going to be too controversial for the Inquisition.

Despite his caution in publishing, Descartes did manage to contribute to both science and mathematics: he basically established the field of analytic geometry, and we still speak of the *Cartesian* coordinates system. He discovered the law of refraction and contributed to cosmology and to the discussion of what "forces" in physics really are. But above all, Descartes is relevant to our story for two valiant attempts (both largely failed) to contribute to science's modus operandi: his "radical doubt" and the articulation of his "method." See, while Descartes—like Bacon—was attempting to overturn the Aristotelian-Scholastic way of thinking, he was convinced—unlike Bacon—that Aristotle got it right when he declared the primacy of deduction. For Descartes, the issue was clear: mathematics is the most successful of sciences, and it is entirely based on deduction. Ergo, if we want successful natural sciences we need to find a way to use deduction as the foundation of science itself.[7]

Descartes went about pursuing his goal in a highly original and daring manner, if one that was doomed to failure. The reason this is important for our discussion is because Descartes' attempt was arguably the last of its kind before the complete separation of philosophy and science, and it is his failure that highlights the difference between the two disciplines that still characterizes them today. The first of the two components of Descartes' strategy, his "radical doubt," is often encapsulated in his famous phrase *Cogito, ergo sum* (I think, therefore I am). The philosopher imagined that no matter how solid his knowledge of things of the world may appear to be, it is always possible to conceive of an evil demon who could trick him into believing things that did not exist. Clearly, human senses can be misleading, which is plainly shown by the kind of dream that feels real while it is happening or by phenomena like mirages. Even human reasoning is faulty, again as shown by the fact that we can be absolutely convinced of the soundness of an argument only to be ruthlessly shown wrong by someone who has looked at it more carefully or from a different angle. What then? Descartes concluded that there is one, and only one, thing of which he could be absolutely sure: that he existed. The reason for this was that he was capable of thinking and aware of the fact that he was thinking. Not even a supremely clever demon could conjure that kind of trick and make someone who doesn't exist think that he is thinking (indeed, the very notion seems self-contradictory and therefore absurd).

Having achieved this milestone through his exercise of radical doubt, Descartes then set out to re-derive all philosophical (and scientific) knowledge from that firm foundation using his "method." Here I paraphrase the steps Descartes proposed to undertake in his famous, and self-explanatorily titled, *Discourse on the Method of Rightly Conducting Reason*, published in 1637:

1. Accept as true only what cannot be doubted.
2. Subdivide every problem into manageable bits.
3. Tackle the easy stuff first, get to the complicated things later.
4. Review frequently, so that the whole edifice continuously sustains scrutiny.

We have already talked about the first step—it's the one that resulted in the *cogito*. It is also the most problematic, because it turns out that we cannot be certain of anything *other* than the fact that we exist. This is not to say that all knowledge is relative, and your opinion on, say, black holes is as good as that of cosmologist Stephen Hawking. Rather, knowledge of the real world comes in degrees, and it is prudent to treat it in probabilistic terms. When scientists say, for instance, that AIDS *is* caused by the human immunodeficiency virus, they mean this to be a statement that comes with a "very high likelihood" sticker attached to it. It is possible to doubt it, and even to reject it, but the more probable the statement is considered to be (because of evidence in its favor), the higher is the burden of proof on the skeptics who doubt that HIV and AIDS are connected. What makes AIDS skeptics obtuse (willfully or not), charlatans, or, worse, criminals (when their attitudes cost human lives) is not that they are doubting a scientific "truth," but rather that they are making an extraordinary claim without putting up extraordinary evidence to back it up.

Interestingly though, the remaining three steps on Descartes' list are, in some fashion or another, accepted parts of how science works today (even if they by no means exhaust the complex set of activities that go under the rubric of "scientific method"), although they were quite revolutionary at the time. As to the second point on Descartes' list, scientists as a matter of course do try to break down problems into manageable bits. For instance, one would be ill advised to begin a research program aimed directly at the big question of how life originated because, well, it is a *big* question. Instead, various laboratories may identify smaller bits of the issue, such as how specific organic molecules appeared on the prebiotic earth, how the first cell membranes were formed, how nucleic acid (DNA and RNA) replication came about, and so forth, and begin by tackling those first.[8] This is rather commonsensical: when we want to achieve something big in our lives, we don't do it all at once either, but rather we start with the small stuff in an attempt to reach the larger goal, even if that goal evolves over time. If I want to write a book, I don't start from the beginning and write straight through the end. Rather, I prepare an array of topics to be grouped in chapters; I do research on a small number of subtopics at a time; I write individual chunks of individual chapters, likely not according to a smooth schedule, nor certainly in the order in which they will eventually appear in the book. Through all of this, however, the writer as well as the scientist ought to have a (often evolving) sense of where they are going, of the overall goal.

Descartes' third step also makes sense, and is in fact intimately connected to the previous one: dividing large problems into manageable bits will rarely result in bits of equal complexity or difficulty, and the scientist will be wise to tackle the easy ones first. This is just like trying to solve a complicated puzzle with tens of thousands of pieces: not only does one proceed more quickly (and gain a psychological boost) by placing the easy bits first, but the whole picture becomes increasingly constrained so that pieces that would have been very difficult to place from scratch may actually fall into place with relative ease.

This brings us to the fourth and last item on Descartes' list: the constant reexamination of the current status of the puzzle to make sure we are still on the right trajectory. As we shall see in the next chapter, scientists do commit blunders, and science has from time to time found itself in a blind alley from which it had to retreat before rejoining the main road. Again, the analogy with puzzles is apt: at some point, the player may realize that an early mistake was made and may have to retrace his steps or, in the worst-case scenario, actually start all over again. But notice the positive side of this way of proceeding: every new piece of the puzzle that is successfully placed is at the same time a test for the entire construction, and the more tests the construction stands, the more confident we grow that we may have gotten it right. This analogy also makes it clear why science does not *need* Descartes' first step (fortunately, because it is impossible to achieve!): scientific theories are tested every time someone makes an observation or conducts an experiment, so it is misleading to think of science as an edifice built on foundations. Rather, scientific knowledge is more like a web.[9] The difference couldn't be more crucial: a tall edifice can collapse all at once if the foundations on which it was built turn out to be shaky, but a web can be torn in several parts without causing the collapse of the whole. The damaged threads can be patiently replaced and reconnected with the rest, and the whole web can become stronger and more intricate. This is exactly the way scientists began to work after Descartes, starting with the example set by our next two characters: Galileo and Newton.

The Italian Galileo Galilei consciously adopted the double title of "philosopher" and "mathematician" when he left the University of Padua to return to Florence in 1611, under the patronage of the powerful Medici family.[10] The term "scientist," by which later generations categorize him, was not in common parlance yet (it was introduced in the 1830s, in direct analogy with the term "artist"). Galileo, who lived at the same time as Descartes, was the third in our series of transitional figures who consciously set out to overturn the Aristotelian view of the world—and he got into deep trouble with the Catholic Church for his efforts.

In 1616 the Santo Uffizio (Saintly Office) of the church declared the Copernican theory of a heliocentric solar system "absurd" and allowed its presentation only in the form of a mere hypothesis, but not as a fact.[11] Galileo had written unpublished (but widely circulated) letters sympathetic to the Copernican theory, which began to attract unwelcome attention to his work. Undeterred, and probably counting on his good relations with the new pope, Urban VIII, Galileo continued to publish daring defenses of the Copernican system, culminating in his *Dialogue Concerning the Two Chief World Systems*, published in 1632. In a shrewd move, Galileo went to Rome to obtain a papal certification that the book was in conformity with the edict of 1616 (which it wasn't, and Galileo knew it). A busy Urban skimmed through the manuscript and gave the imprimatur (essentially the authorization for publication) only to regret it soon after (the pope even thought that one of the characters of the dialogue, the inept Simplicius, was meant as a caricature of himself—though this very likely was not Galileo's intention). Be that as it may, the Florentine philosopher-scientist was brought to trial in Rome in 1633 (when he was already sixty-nine). He was swiftly found guilty and condemned to a series of humiliations, including the listing of the *Dialogue* on the Index of Prohibited Books, an agreement to officially renounce the Copernican theory, a period in prison (later commuted to house arrest in Tuscany), and the recitation for three years, once a week, of seven penitential psalms (which was delegated to his daughter Maria Celeste, a nun). All in all, much better than the outcome of the trial of Giordano Bruno only three decades earlier, but still, a sad ending to a prestigious life and career.

Indeed, it took an inordinate amount of time for Galileo's reputation to be restored with the church (although the rest of the world quickly came to think of him as a genius and a martyr to obscurantist forces). We had to wait until 1846 before books in support of the Copernican theory were taken off the index and as late as 1968 for Pope Paul VI to begin the revision of the trial against Galileo. Interestingly, however, Cardinal Joseph Ratzinger (later Pope Benedict XVI) raised questions against such revision on the ground that the trial had allegedly been "reasonable and just" and citing in his favor radical philosopher Paul Feyerabend, on whose lunacy I will comment in detail later in the book. Finally, it was under Pope John Paul II, in 1992, that Galileo (but not Bruno) was officially pardoned. It is hard to conjure a more obvious demonstration of the stubbornness and irrelevance of religious authorities when it comes to their relationship with science.

But I digress. The reason Galileo is important to our story is because not only was he yet another anti-Aristotelian rabble rouser, but more precisely because of the way he went about criticizing the Greek master. We have seen that both Bacon and Descartes took Aristotle to task on philosophical grounds, which places both of them firmly on one side of the philosophy-science border. Galileo, instead, planted himself equally firmly on the science side. His assault on Aristotle started in 1590, with *On Motion*, and continued in 1610 with the publication of *The Starry Messenger*, in which Galileo presented his early discoveries made through the telescope.[12] Galileo's strategy was to show by empirical example that the distinction made by Aristotle between terrestrial and celestial realms was illusory. For instance, he described the mountains of the moon by explicitly drawing the parallel with those in Bohemia. Later on, Galileo showed the existence of dark spots on the surface of the sun and argued that they demonstrated that our star rotates on its axis, just like the earth. He also discovered the existence of phases of the planet Venus, analogous to those of the moon. The implication was that there was no sharp difference between the sublunar and "corrupt" world and the heavenly and allegedly perfect one. From there it is a short step to concluding that there is only one kind of matter (as opposed to the classical Greek distinction of four elements: earth, water, fire, and wind) and even only one kind of motion—a radical notion that came to full fruition later with the work of Newton.

Perhaps the most elegant and devastating critique of the Aristotelian conception of physics was made by Galileo through a thought experiment:[13] Aristotle had claimed that bodies fall at a speed that depends on their weight, the heavier bodies faster than the lighter ones. Galileo therefore invited his readers to consider a situation in which a heavier and a lighter body were joined by a cord and were falling together. According to Aristotelian physics, the heavier one should drag the lighter one down faster so that the combined body would have a higher speed than the lighter by itself. Then again, by the same Aristotelian theory, the lighter body would be expected to *slow down* the heavier one, so that the combined bodies would fall at a slower speed than the heavy body by itself. The punch line is that the first prediction is logically contradictory to the second one: one cannot be true without the other one being false, and yet Aristotle would have to admit that both stem out of his conception of bodies and speed. Ergo, Aristotle's conception of bodies and speed must be wrong. One can imagine the ancient Greek nodding approvingly at the genius of the Italian scientist and admitting defeat. Although common lore wants it that Galileo, being a good scientist, actually performed the experiment by dropping different weights from the tower in Pisa, there is much historical doubt that he actually did. However, Neil Armstrong carried out the physical experiment (by that point a demonstration, since no one doubted the outcome) in spectacular fashion on the moon, during the *Apollo 11* mission of 1969.[14]

Our next character in this brief panorama lived after Bacon, Descartes, and Galileo (he was born the year Galileo died) and yet was also clearly a transitional figure, albeit one actually more pertinent to the theme of this book, the sometimes fuzzy distinction between science and pseudoscience. We turn now to Sir Isaac Newton.

Newton is hardly a likable character. Despite being one of the undisputed scientific geniuses of all times, he was apparently also a rather nasty man. Most famously, he got involved in a long-lasting dispute with the philosopher Gottried Leibniz over who first invented calculus (the idea that—among other things—solved Zeno's paradox, encountered in the last chapter). The reality is probably that Newton arrived at the idea before Leibniz, though the latter published first; in any case, the two could make an equal claim to the invention because they formulated their ideas independently of each other. Nonetheless, Newton went so far as to personally appoint the members of a commission of the Royal Society to settle the matter, with the predictable result that the commission ruled in Newton's favor. Even more outrageously, Newton probably penned the final report of the commission himself, though he did not sign it. He harbored deep rancor against Leibniz, which continued after the latter's death. And this wasn't the only instance of spite by Sir Isaac: when Robert Hooke criticized some of his ideas on optics, Newton considered him a foe until the end of his life. His famous phrase "If I have seen further, it is by standing on the shoulders of giants," often quoted as a sign of humility by a great scientist, was nothing of the kind: it was more likely a nasty jab at Hooke, who was a very short man. Finally, when Newton became president of the Royal Society (the position that allowed him to indirectly adjudicate his own dispute with Leibniz), he attempted to steal royal astronomer John Flamsteed's catalog of observations.

Then again, all of this simply goes to exemplify the general conclusion that personal character has little to do with scientific (or, for that matter, artistic) genius. Newton is justly famous for his many fundamental contributions to science, beginning with the formulation of the universal laws of motion that completed the work initiated by Galileo and by Johannes Kepler (and which, incidentally, finally removed any lingering doubts about the basic correctness of the Copernican system). Newton's ideas about absolute space and time defined physics and the way we thought of the fabric of the universe until Einstein's theory of relativity centuries later. And Newton's laws of motion, though only a rough approximation when compared to the relativistic ones, are still so useful that they are employed for most practical purposes (e.g., in calculating the trajectory of a space probe aimed at Mars). His *Philosophiae Naturalis Principia Mathematica*, published in 1687, is one of the milestones of the scientific endeavor, and his theory of gravitation alone would have been sufficient to anoint him as one of the greatest thinkers in history.[15]

The complexity of Newton's figure is particularly pertinent to our discussion because, as the British economist John Maynard Keynes once said, "Newton was not the first of the age of reason. He was the last of the magicians, the last of the Babylonians and Sumerians, the last great mind that looked out on the visible and intellectual world with the same eyes as those who began to build our intellectual inheritance rather less than 10,000 years ago." What was Keynes talking about? Newton's interest for alchemy and the Bible, the latter of which is still quoted by modern creationists as "proof" that great scientists believe in a literal reading of their scriptures.[16] Newton's fascination with alchemy was catalyzed by his colleague and patron at Cambridge, the philosopher Henry More. By some accounts, Newton spent more of his active life on alchemy than on physics, and he probably paid a dire price for it, suffering at least one nervous breakdown and hedging close to madness, likely as a result of inhaling mercury during his alchemic experiments. Apparently he placed alchemy on the same level as his other scientific interests, a position that would be considered stunningly ridiculous by modern standards.

Newton's interest in a literal reading of the Bible is no less surprising in someone (justly) considered one of the fathers of modern science. He wrote several essays in which he attempted to extract scientific information from the scriptures and was convinced that the end of the world would occur in 2060 (well, at least that is an empirically verifiable prediction: mark your calendars!). Then again, even in this he was a strange mixture of past and future attitudes: his study of the Bible led him to conclude that the idea of a Trinity had been added by later writers and that God was best thought of as a cosmic clockmaker who began everything by laying down the laws of physics. This idea is somewhat of a forerunner of a position popular during the Enlightenment (and embraced by several founding fathers of the United States) known as "deism." In deism, God does not intervene regularly in human affairs (so prayers and miracles are out), but instead planned the unfolding of cosmic events as an all-powerful engineer.

It is hard to reconcile all of this with Newton's own exposition of his scientific method, which boils down to the following:

- Never entertain unnecessary explanations.
- Causes are proportional to effects.
- Qualities that are the same for different bodies must be considered as universal.
- Conclusions arrived at by induction based on experiments must be considered true until otherwise demonstrated.

Difficult to imagine where alchemy or Bible studies would fit into the above list, but the fact that a brilliant mind like Newton's couldn't see the inherent contradictions in what he was doing ought to be a cautionary tale for anyone who believes that thinking critically is an easy task. Nonetheless, Newton's scientific contributions eventually overshadowed his penchant for mysticism to the extent that today he is considered one of the founding figures of science. A good reason for this is that science, unlike alchemy and biblical literalism, keeps making progress, with thousands of scientists over centuries contributing to improve on what Newton started.

Despite the inconsistencies we have seen in the work of these four figures, thanks to the efforts of Newton, Bacon, Descartes, Galileo, and several others, science was now well on its way to a complete separation from both religion and philosophy, for the first time in human history. Before concluding the chapter by looking at the current, more nuanced and complex relationship between science and philosophy, however, we need to turn briefly to yet another crucial figure: Charles Darwin. Darwin, besides his well-know contributions to biology as a scientist, was also very concerned about the soundness of his philosophical methods. His was a mature preoccupation that highlights the independence that science had gained from philosophy by the Victorian era, and it is much more in line with Gaileo's scientific pragmatism than with Descartes' metaphysical preoccupations.

According to philosopher David Hull, "evolutionary theory seems capable of offending almost everyone,"[17] something that hasn't changed since the publication of Darwin's *Origin of Species* in 1859. Of course, some of Darwin's contemporaries also raised genuinely scientific objections to the theories of common descent and natural selection, but such objections were quickly overcome, and evolution was soon considered as controversial in biology as quantum mechanics is in modern physics (i.e., not at all). Nonetheless, in reading Hull's *Darwin and His Critics* one discovers that an entirely different sort of attack caused quite a bit of trouble for Darwin and his fledging theory, an attack that tells us something about the conceptual history of science. Darwin got himself caught in the middle of the great induction debate in the then nascent field of philosophy of science. Moreover, he seemed to be on the losing side.

The great induction debate unfolded between two of the major British philosophers of Darwin's time, John Stuart Mill and William Whewell. Both were attempting to improve on Francis Bacon's notion of induction. As we have seen, Bacon wanted to establish the scientific method on firmer foundations than what he thought Aristotle had achieved. Bacon, unlike Aristotle, favored the inductive over the deductive method.[18] But we also noted in the last chapter that Scottish philosopher David Hume had pointed out that this wouldn't do either, because induction amounts to a questionable extrapolation from what is known (the observations) to what is unknown (what one predicts), with no independent assurance that new phenomena in the world will emerge from the same causes that produced the old ones.

Both Mill and Whewell therefore tried to improve on Bacon in light of Hume's discussion (and people argue that philosophy doesn't make progress!). Mill thought that scientists could use two kinds of induction, which would mutually reinforce each other. *Enumerative induction* is a strengthened version of Bacon's in which the process of extrapolation is made sound by the law of universal causation (i.e., the idea that all phenomena have causes and that it is logical to attribute similar causes to similar phenomena). *Eliminative induction* is an operation by which the causes of natural phenomena are discovered by successive elimination of unsuitable alternatives, based on tests as stringent as can be devised (this is reminiscent of the idea of "strong inference" that we encountered earlier in the book).

Whewell would have none of it, because he thought that scientific investigation had to start with hypotheses, not observations, because hypotheses have the value of guiding one's inquiry (i.e., they tell us where and how to observe nature). When a hypothesis is confronted with the data, one knows if it is true, according to Whewell, because of what he called "consilience": "The Consilience of Inductions takes place when an Induction, obtained from one class of facts, coincides with an Induction obtained from another different class. Thus Consilience is a test of the truth of the Theory in which it occurs." Whewell also (confusingly) retained the term "induction" for his view, probably because of the patina of prestige lent to the term by Bacon.

There were problems with both views. Mill did not like the idea of judging the truth of hypotheses by a consilience-type cross-test because of the well-known issue that more than one hypothesis could fit the observations equally well (so the consilience process could end up jumping to the wrong conclusion). Then again, neither enumerative nor eliminative induction are logical guarantors of truth because they are themselves based on certain assumptions about the world. (It turns out that there is no logical guarantor of truth about nature, if one does not count mathematical proofs. Scientists just have to do with more or less likely versions of "maybe.")

What does have all of this to do with Darwin? The problem is that both Mill and Whewell—regardless of their internal disagreements—saw Darwin's work in the *Origin* as largely based on deduction and therefore as bad science! Darwin had, starting from a few observations, jumped to the generalization of natural selection and had then deduced a series of consequences, which he spent twenty years trying to back up with additional observations. For both philosophers this did not qualify as induction and was therefore simply unscientific. As Hull puts it, the philosophers were "equally adamant in their conviction that the *Origin of Species* was just one mass of conjecture."

It turns out, however, that the philosophers were wrong and the scientist was on the mark. Ironically, an analysis of Darwin's work and of how he himself describes it reveals that his method was actually very close to the one proposed by Whewell (though in clear opposition to Mill's). In Darwin's own words: "How odd it is that anyone should not see that all observation must be for or against some view if it is to be of any service!" That is, Darwin was applying precisely Whewell's method of consilience, though this was not at all clear to Whewell himself. The reason this episode is important is because it contributed to establishing biology as independent not only from philosophy but also from what was then (and still is, in some quarters) a sense that "science" must be synonymous with the way physicists like Descartes, Galileo, and Newton worked. This was one the first historical hints that "science" actually refers to a rather heterogeneous family of activities, not to a single and universal method.

The Complex Relationship between Philosophy and Science

As we have seen in the story so far, the relationship between science and philosophy has always been tormented. Science slowly evolved from "natural philosophy" during the sixteenth and seventeenth centuries, thanks to the work of people who thought of themselves as philosophers and most of whom we consider scientists. As in any parent-offspring relationship, things can get acrimonious, with the offspring staking out its territory while denying the parent's relevance or contribution, and the parent having a difficult time letting go of the now mature and independent progeny.

In the remainder of this chapter I will sketch the current state of the science-philosophy borderlands, sometimes unofficially referred to as "sci-phi," because both fields are instrumental to our quest for understanding where science ends and pseudoscience begins. We will start by looking at what modern philosophy of science actually consists of, exploring some instances where it can directly contribute to science and others where it is best thought as an independent field of study. We will then examine examples of how philosophers can be useful to scientists not just within the narrow confines of scientific research itself, but in the broader quest for a better understanding of science on the part of the general public. We will see that "sci-phi" constitutes a formidable intellectual weapon in the defense against pseudoscience and the quest for more widespread appreciation and practice of critical thinking.

Just as philosophy is seen rather skeptically by the general public, not all scientists agree that it is a useful pursuit. Nobel physicist Steven Weinberg, for instance, took the rather unusual step of writing a whole essay entitled "Against Philosophy."[19] In it, Weinberg argues that not only is philosophy not useful to science, but in some instances it can be positively harmful. The example he provides is the alleged slow acceptance of quantum mechanics—by most accounts the single most successful scientific theory of all times—because so many scientists at the start of the twentieth century, beginning with Einstein, were unduly fascinated by a philosophical school known as logical positivism.

Logical positivism is a now largely abandoned philosophical position—originally associated with the so-called Vienna Circle of intellectuals—that took a rather narrow view of what counts as science. Most famously, logical positivists thought that science had no business dealing with "unobservables," that is, with postulating the existence of entities that cannot be subjected to experimental tests. Quantum mechanics is rife with such unobservables, such as quarks, electrons, forces, and so on, and positivists were indeed highly skeptical of the whole affair, which smelled too much of metaphysics (a very bad word, in their vocabulary). It is also true that some scientists were initially rather uncomfortable with the wildest implications of quantum mechanics (as in Einstein's famous quip that "God doesn't play dice," referring to the fundamental randomness of quantum effects) and resisted them while searching for alternative interpretations of the theory.

Nonetheless, I seriously doubt that one can blame a probably nonexistent slowing down of turn-of-the-century physics on philosophy. First of all, philosophy has simply never had, in recent memory anyway, much of a sway with scientists. True, Einstein was influenced by the positivism of philosopher Ernst Mach, and we have seen above Darwin's struggle with Whewell and Mill. But these are rather exceptional and increasingly rare cases. This is in part because scientists tend to be interested in whatever methods maximize discovery, not in philosophical issues about metaphysical truths. Second, although it is of course impossible to rewind the tape of history and experiment with possible alternative historical sequences, quantum mechanics was actually accepted very quickly by the scientific community, especially when one considers its radical departure from any previous physical theory. For example, quantum mechanics introduced the concept that certain properties of matter cannot be measured simultaneously (the so-called Heisenberg indeterminacy principle), not because our instruments may not be up to the task, but as a consequence of how the universe itself works!

A diametrically opposite view to Weinberg's is the one expressed by Daniel Dennett (perhaps not surprisingly a philosopher) in his book *Darwin's Dangerous Idea*: "There is no such thing as philosophy-free science; there is only science whose philosophical baggage is taken on board without examination."[20] This will strike most scientists as preposterously arrogant, but a moment's reflection shows that Dennett, of course, is right. For example, as we have seen in some detail, scientific practice requires the assumption of naturalism, that is, the idea that natural phenomena are, well, natural, and therefore scientists do not need to invoke the supernatural to explain them. In fact, we noted that scientists themselves invoke naturalism as a postulate of science whenever they need to make the (convincing) argument that the so-called theory of intelligent design is not science *because* of its supernatural basis. The important thing to realize is that in the context of our discussion naturalism is not an empirically verifiable position, and it is therefore by definition outside of science itself. If science is about anything at all, it is about empirically verifiable statements concerning the world.

Attitudes such as Weinberg's are largely the result of intellectual arrogance. I am convinced that such arrogance hurts science, and it certainly doesn't help bridge what C. P. Snow famously referred to as the divide between "the two cultures."[21] Let's therefore try to move beyond simplistic academic posturing and see what the proper domains of philosophy of science actually are, and where they do or do not intersect with the practice of science as understood since Galileo and Newton. Broadly speaking, philosophy of science deals with three general areas of inquiry: the nature of science, the conceptual and methodological analysis of science, and science criticism.

Most scientists, if they are familiar with philosophy at all, have some acquaintance with philosophical studies on the nature of science. Names like Karl Popper and Thomas Kuhn even make it into the occasional introductory science textbook, and one can argue that the ideas for which these two philosophers are respectively most famous, falsificationism and paradigm shifts, are among the few concepts in modern philosophy of science that are ever mentioned in the halls of science departments. Popper and falsificationism are representative of a somewhat prescriptive streak in philosophy of science, that is, of a tradition of philosophers actually telling scientists how they ought to carry out their work. As we have seen, Popper was motivated by the so-called demarcation problem—the difficulty in distinguishing science from non- and pseudoscience—the very issue that is central to this book. He was also bothered by Hume's problem of induction, the idea—which we have already explored—that science is based on induction, and yet the only reason we have to trust induction is because it worked in the past (which is itself a form of induction, making the whole thing perilously close to being circular).

It is rather ironic that many science textbooks have essentially adopted Popper's view of science as an enterprise dealing in falsificationism, with many scientists actually defining science in Popperian terms. We have seen at the very onset of this book that Popperian falsificationism has long been superseded in philosophy of science on the grounds that science just doesn't work that way. Think, for instance, of the fact that the original Copernican theory did not actually fit the data very well at all, and yet it was not rejected as "falsified." Rather, scientists gave it some time to develop because it seemed a promising approach, and sure enough, Galileo, Kepler, and Newton all improved on both the theory and the quality of available data, and we now speak of the success of the Copernican revolution. We will see later in this book why Kuhn's ideas about how science works are also far from being unproblematic, but for now it is sufficient to note that he was not attempting to prescribe how scientists ought to work, just to describe how they do, in fact, work. The important point is that while scientists are aware of both the prescriptive and descriptive streaks in the philosophical study of the nature of science, they have apparently accepted notions from philosophy that have been superseded, or at least have been criticized as highly problematic, without much evidence that more recent formulations by current philosophers have even made it onto the scientist's radar screen. This should be rather embarrassing (imagine the scandal if a philosopher claimed that Darwin's book is the latest news in evolutionary biology or that Newton's theory is the current view in the physics of motion . . .), but it isn't, because scientists simply don't seem to think that philosophy is relevant at all.

The second major area of inquiry in philosophy of science is concerned with the conceptual and methodological analysis of what scientists actually do. David Hume was among the first to take this approach, wondering for example about our concept of causality (his analysis, still surprisingly challenging today, was not very encouraging, as he concluded that we do not really know what we mean when we talk about "causes"). More recently, philosophers have been sharply critical of the very popular field of "evolutionary psychology," and we have seen the reasons for such criticism earlier in this book.

I refer to the third and last major type of philosophy of science as science criticism (often called "science studies," though the latter has a rather controversial meaning attached to it), and it addresses the complex interface between science and society that concerns us all. For example, philosophical issues surrounding the so-called nature-nurture debate are relevant to the uses and, more importantly, the misuses of genetic medicine.[22] Here the philosopher becomes a critic not just of how the science is being conducted and its findings interpreted, but primarily of how such findings are understood by the public and used to guide social policies.

What is a scientist to do with all this? Scientists may largely and safely ignore what philosophers say about the first area, how science does or should work in broad terms—after all, scientists want to do science, not to think about how it's done. However, they do have a responsibility to at least update their understanding of philosophy when it comes to writing science textbooks or teaching the nature of science in their classrooms. Also, it should go without saying (but in fact it needs to be said) that philosophers have the intellectual right to pursue whatever sort of inquiry they wish to pursue into the nature of science, without having to justify themselves to scientists by defending the "utility" (implicitly, to science) of what they do.

But when we move to the second and third areas of philosophical inquiry sketched above we come closer to the sci-phi borderlands, to the point that, in some cases, philosophy may be thought of as "the continuation of science by other means."[23] Indeed, in certain areas of theoretical science it is sometimes difficult to tell whether a paper is written by a scientist or by a philosopher without directly checking the author's institutional affiliation. Here the word "theory" takes on its original and broader meaning of formulation of concepts, not just mathematical treatment (although there are examples of philosophers engaging in the latter as well). What makes this blurred line between philosophy and science interesting is that the two disciplines bring different backgrounds and approaches to the study of the same issues—that is, this isn't just a matter of science-envy by philosophers (or the even rarer phenomenon of philosophy-envy by scientists).

There are many published examples of interaction between science (particularly biology) and philosophy when it comes to squarely scientific questions, ranging from how organisms construct their ecological niches to the meaning of various concepts of biological species, just to mention a couple of those that have been vigorously debated in recent years. What interests us here, however, is the potential for fruitful interactions between science and philosophy when it comes to a joint defense against the assault from pseudoscientific quarters. I will sketch a simple example below as an illustration of how the methods of scientific and philosophical inquiries can be complementary in debunking pseudoscientific claims.

Consider the oft-made contention that people have "out-of-body" experiences in which, under certain conditions (for example during a surgical operation), a subject recalls having found himself somehow outside of his own body. He observes the scene from a different point of view and even recalls some of the things that were said or done despite the fact that he was under sedation at the time.[24] Now the classical, scientific approach to analyzing this sort of claim may begin, for example, by considering the physical circumstances of the alleged event, such as the details of the operating room, the conditions of the patient, etc. A scientifically minded investigator may then proceed to look for corroborating details supporting the subject's version of the story: was he really able to hear what the doctors were saying? Did he actually somehow observe specific events that occurred in the room? A scientist may then propose possible alternative explanations for the facts so gathered—for example, that the experience was caused by a side effect of the anesthetics used in preparation for the operation, and so forth.

This modus operandi is typical of science in general, not just when applied to claims of the paranormal. The idea is that one works within certain assumptions—for example, that there is no conscious attempt to deceive the investigator—just as in the case of normal scientific research, where fraud is a hypothesis of last resort. The investigator then focuses on the details of the alleged experience, attempting to see if they do or do not square with the available evidence, essentially playing the role of an investigative detective—an analogy often brought up to explain the nature of everyday scientific research.

A philosopher, however, would approach the same problem in a different fashion. In particular, she would focus on the broad picture, on the validity of background assumptions, and on the internal coherence of the claims under investigation. For example, a question that might be posed is what, exactly, does it mean to "see" one's body from outside? To the best of our knowledge, seeing is something we do with a complex bodily apparatus that includes not just eyes but also a connected brain that is capable of interpreting light signals. But if one is disembodied, how would "vision" work? Moreover, subjects who claim to have undergone out-of-body experiences usually talk as if their vision were of the same kind as ordinary vision, that is, with a limited comprised angle. But if vision were somehow possible outside of the constraints imposed by biological structures, why wouldn't people be able to see the full 360° of their surroundings? These sorts of questions would complement, and perhaps even aid, the scientific-type approach, resulting in the conclusion that the experience must have been more psychological than physical, since not only doesn't it square with verifiable details (science), but it is internally incoherent (philosophy)—much like a dream. The point of this exercise is hardly to convince the true believer, but this kind of dual approach could be used to teach students and the general public about the nature of both science and philosophy, and at the same time one might be able to heighten the critical sense of people when they read about alleged paranormal phenomena.

There is much talk (and usually little action) on university campuses about interdisciplinary teaching and scholarship (the concept is not even addressed, for a variety of reasons, at the precollege level, where it would be likely to have the most impact). A more serious exploration of the borderlands between science and philosophy offers a fertile ground for the kind of interdisciplinarity that can move forward both fields as well as have a positive effect on students and the general public. Again, the idea is not that philosophy will answer scientific questions, as Weinberg demanded in his essay "against philosophy." For that sort of task we have science, and it works very well. Nor am I leaning toward a model under which science will somehow "reduce" philosophy to biology or chemistry, as has been suggested in a rather simplistic manner by biologist E. O. Wilson.[25]

Rather, the model is one of partially overlapping, but largely distinct spheres of influence. For one thing, the aims and methods of the two disciplines are now different as a result of their diverging conceptual history, which we have traced in the last two chapters: philosophers tend to be interested in posing questions and in analyzing concepts while scientists are inclined toward pragmatic solutions (heuristics) that maximize discovery and concentrate on details of data gathering and analysis. More broadly, philosophy is concerned with a range of topics that are often outside the scope of empirically based approaches, like metaphysics, which scientists ought to be very happy to leave to someone else to deal with.

Given all of the above, why then is it so difficult for scientists and philosophers to just get along? The reasons are various and complex and include a certain degree of diffidence toward an unfamiliar field (nobody likes the sudden feeling of being a novice, especially if one has spent most of his life being an expert in a recognized area of inquiry). This is perhaps augmented by a dose of intellectual arrogance, as in the oft-heard remark that philosophy is "just armchair speculation" (so, technically, is computer modeling or mathematical biology or a lot of engineering, though they are never referred to with that dismissive appellative). The whole postmodern wave in late twentieth century philosophy, which we will discuss in a couple of chapters, hasn't helped either by producing egregious examples of nonsense on stilts, such as philosopher Paul Feyerabend's contention that astrology and magic are as legitimate as science, or Michel Foucault's overly simplistic statement that there are no universal truths, ever. In the end, it is up to individual philosophers and scientists to seriously consider when they can march on, largely unconcerned with what the other side has to say, or when and to what extent their respective fields can benefit from each other's influence. Both philosophy and science have come a long way since the time of the pre-Socratics, but their combined use still represents the most formidable intellectual weapon against nonsense that humanity has ever devised. Aristotle would have been pleased.

CHAPTER TEN

The Science Wars I

Do We Trust Science Too Much?

I have not the smallest molecule of faith in aerial navigation other than ballooning, or of the expectation of good results from any of the trials we hear of.—Lord Kelvin, 1896

Science is the great antidote to the poison of enthusiasm and superstition.—Adam Smith

"The natural world has a small or non-existent role in the construction of scientific knowledge," wrote sociologist Harry Collins in the introduction to a collection of works[1] by sociologists of science of the "relativist" persuasion, which he thought brought "substantive" contributions to the field. Collins's is a startling claim, one that runs against most people's intuition that science, on the contrary, has everything to do with the natural world. If gaining knowledge about nature is the whole point of science, how can Collins assert that the two have no relationship with each other? The answer is that Collins belongs to a loosely defined movement in the social sciences and humanities that has been criticizing science on the basis that it is a human activity, not an objective march towards the Truth. This movement is referred to by a variety of terms that actually indicate a family of schools of thought originally rooted in literary criticism and that have subsequently expanded into philosophy and the broader humanities: "postmodernism" and "deconstructionism" are the most common terms (and the two are not exactly equivalent), although for the purposes of this chapter I will actually refer to them under the umbrella of "constructivism," in reference to the basic idea they all share: that scientific knowledge (or any kind of knowledge, for that matter) is socially constructed with little or no input from outside of human conventions.[2]

In the next chapter, I will argue that the extreme constructivist position is naïve and ends up being harmful not only to science, but to society at large. On the other hand, its logical opposite—which will be referred to here as "objectivism"[3] and which will be discussed in this chapter—is also naïve and harmful, though in a quite different fashion. Objectivism in this sense is the position implicitly (and sometimes explicitly) advocated by scientists, many of whom really seem to be convinced that science affords them a "God's-eye" view of the universe, capable of somehow transcending the limitations of other human enterprises and reaching out to grab the Truth (notice the capital *T*).

Of course, there is a third way, to which we will get in due course, and which manages to incorporate the reasonable criticisms mounted by constructivists while retaining a special place for science as a uniquely effective way to gain knowledge about the natural world, pace Collins and colleagues. But before we get to the resolution of the conflict, let us take a brief tour through some of science's big failures (this chapter), as well as examine some of constructivists' major blunders (next chapter), peppering the whole journey with enough basic ideas in philosophy of science to provide us with the tools to understand what the so-called science wars have been, and to some extent still are, all about.

No God's-Eye View

Scientists are not gods, even though one may sometimes have some difficulty making the distinction, judging from the ego that some of them (the scientists, not the gods) display when talking about what they do. It is not uncommon to hear physicists and cosmologists expounding on the possibility of "theories of everything," although what they mean is actually a mathematical solution to a specific problem concerning the conceptual unification of natural forces. Cosmologists such as Stephen Hawking freely talk about having seen "the mind of God" when they come up with a new theory about the distant future of the universe (never mind that so far we do not have a unified theory of forces or that Hawking's initial predictions about the fate of the universe have been proven spectacularly wrong by recent empirical research).[4] Or consider biologist Richard Dawkins, who goes so far as to (mistakenly, as it turns out) claim that science can refute what he calls "the God hypothesis."[5]

The examples above are instances of *scientism*, a term that sounds descriptive but is in fact only used as an insult. The term "scientism" encapsulates the intellectual arrogance of some scientists who think that, given enough time and especially financial resources, science will be able to answer whatever meaningful question we may wish to pose—from a cure for cancer to the elusive equation that will tell us how the laws of nature themselves came about.[6] The fact that scientism is an insult, not a philosophical position that anybody cares officially to defend, is perhaps best shown by the fact that there is no noun associated with it: if one engages in scientism one is "being scientistic," not being a scientist.

The scientistic attitude probably has many roots, including the somewhat simplistic and certainly optimistic representation of science we get from TV documentaries and books for children, which, after all, is how most future scientists first become enamored with their subject (in my case it was the writings of astronomer Carl Sagan that did the trick). A certain amount of oversized ego is also likely to be an ingredient for becoming a scientist. The realities of a scientific career are such that one constantly confronts criticism and rejection from one's peers, be that in person at professional meetings or—more insidiously—anonymously in the course of peer reviews of one's grant proposals and technical papers. Unless your ego is large enough to take a significant amount of battering over a span of many years, you will not even finish graduate school.

Yet, I think a major reason for the prevalence of a scientistic attitude among scientists is the equally widespread ignorance of, even contempt for, philosophy. We have already seen that physicist Steven Weinberg devoted an entire chapter of his *Dreams of a Final Theory* to philosophy bashing and unapologetically entitled it "Against Philosophy." Richard Dawkins, who proposed an "argument from improbability" against the existence of God, fails to see that the argument is not scientific (although it is informed by the available scientific worldview), but philosophical: Dawkins rejects "the God Hypothesis" because it fails Occam's razor, the epistemological rule, named after William of Occam (a philosopher) in the fourteenth century, that one should not invoke more explanatory principles than are strictly required by the evidence. In the case of the universe, we can explain much of it by invoking laws of nature, and therefore there is no need for an intelligent designer.

Indeed, the entire scientistic attitude rests on a specific philosophical theory known as the "correspondence theory of truth." In philosophical circles, the correspondence theory of truth has been largely superseded by more sophisticated epistemological positions, although many scientists (and, of course, a large part of the public) are blissfully unaware of this.[7] Epistemology, of course, is the field of philosophy that deals with how we acquire (or fail to acquire) knowledge and therefore arrive at truth. Ever since Plato, knowledge has been thought of as "justified true belief." Let us unpack this latter notion first and then return to the correspondence theory of truth. In the dialogue *Meno*, Plato has Socrates engaged in a discussion of what counts as knowledge, in which Socrates makes an important distinction between mere "true belief" and actual "knowledge." Consider this example: you may have heard, and sincerely believe, that the earth is not flat. As it happens, scientists think that your belief is, in fact, correct, that is, it is *true*. However, according to Socrates, that belief—even if true—does not constitute knowledge on your part unless you can provide an *account* of why one should conclude that the earth is not flat. For instance, you could explain that the hypothesis that the earth is flat is contradicted by the observation of a round shadow projected by our planet on the moon during lunar eclipses; or by the fact that a ship leaving port will disappear gradually from the horizon; or by the fact that one can sail westward and come back to the point of origin after having circumnavigated the planet. In other words, you *know* that the earth isn't flat, as opposed to simply believing the notion, if you can justify your belief. Hence, knowledge is a belief that is both true and justifiable by its holder.

What does any of this have to do with science and its accumulation of knowledge about the natural world? Scientists do hold certain beliefs about the world (e.g., that there are billions of stars in our galaxy); they can also provide justifications for such beliefs (they have scanned the skies with powerful telescopes, counted stars in certain regions, then used statistical techniques to extrapolate and get an estimate of the total number of stars). But are their beliefs true? The correspondence theory of truth says that the way to determine whether something is true or not, or the degree to which it is true (truth does come in degrees in many real circumstances), is to verify that it *corresponds* with how things really are "out there." Ah, but herein lies the problem! How do we know how the world really is? It would seem that this is the sort of pseudoprofound question that philosophers are so often accused of posing, but pondering it for a moment will begin to reveal that there is a real conundrum at play here.

The intuitive response to the question of how we determine the correspondence between what we think is true and what actually is true is the same one that most scientists would give: you just go out there and verify things by experiment and observation. For instance, if I claim that it is true that my friend Phil lives on Long Island, all one has to do is to go to the given address and verify that, indeed, Phil lives on Long Island. What is the big deal? Actually, even in such a simple case things aren't quite that straightforward. The verification in question assumes that just because someone named Phil opens the door at the specified address it follows that he is my friend and that he does live there. But he may be someone else named Phil, or he may be my friend visiting another friend for a short period of time. One then has to embark on a somewhat more sophisticated quest, which might include verifying legal documents (but then you have to assume that the legal system actually works the way you think it works), observing Phil's behavior during a given period of time (assuming that that period of time is actually representative of his recent life conduct), interviewing his friends and colleagues (assuming they are not lying or do not have an interest in keeping you from finding out the truth), and so on.

Of course, my friend Phil *does* in fact live on Long Island (as far as I know), but the point is that this conclusion relies on a variety of sources of evidence (justifications of my belief), each of which can reasonably be questioned. In the case of complex scientific discoveries, such as the number of stars in the Milky Way or the charge of the electron or the reality of biological evolution, the evidence is much more complex, and the number of assumptions that go into reaching a conclusion is proportionally higher. That is why scientific findings should always be considered tentative, open to revision if new evidence comes about.

Anther way to understand the problem is that the correspondence theory of truth essentially assumes that somehow we can step outside of what philosophers call our "epistemic limitations" (e.g., the limits imposed on human knowledge by our biological characteristics) and access a "God's-eye view" of things. While philosophers since Kant have understood that such view is forever inaccessible to human beings, many scientists keep behaving as if science somehow magically allows us to transcend the problem and gain access to the Truth. That is what a scientistic attitude is all about. The ancient Greek philosophers called it *hubris*.

Scientific Blunders and How Science Really Works

Not only can science never in principle reach the Truth because of the untenability of the correspondence theory of truth, but it has also demonstrably blundered in the distant and recent past, sometimes with precisely the sort of dire social consequences that constructionist critics are so worried about. The only reason scientists are by and large unmoved by such instances is because most of them don't "waste" their time studying the history of their own discipline. If they did, they would realize with dismay that every single scientific theory proposed in the past has been shown to be wrong. What makes us think that the story will not continue and that everything we hold true now will not in turn be seen as naïvely wrong by our descendants? Of course, the same history of science can also be read from the diametrically opposite viewpoint, since it equally well illustrates the idea that science is a self-correcting, arguably cumulative enterprise. It may blunder today, but it will likely make up for the mistake a few years or decades down the line.

Either way one looks at it, it is surely instructive to recall the words of Albert Einstein: "There is not the slightest indication that energy will ever be obtainable from the atom."[8] And how about Lord Kelvin, one of the greatest scientists of the nineteenth century, confidently claiming that "X-rays will prove to be a hoax"?[9] And finally, one of my favorites: British astronomer royal Richard Woolley, who in 1956 said, "All this talk about space travel is utter bilge, really."[10] Yuri Gagarin was the first human to orbit the earth just five years later.

A particularly gratifying story of an initial abysmal failure of science, followed by a spectacular redemption, as in all good stories, is the dispute about the age of the earth between Darwin and the above-mentioned Lord Kelvin (aka Sir William Thomson).[11] Thomson was an impressive fellow by all accounts, human as well as scientific. Thomas Huxley, always ready to pick a fight to defend the theory of evolution, thought of Thomson as "a gentle knight" for his courtesy in dealing with fellow scientists. Courteous or not, Thomson greatly upset Darwin because of their disagreement. The problem was this: for Darwin's theory to work, the earth had to be old—really old. Even at the time, no scientist (and certainly not Thomson) took seriously Bishop Ussher's estimate that the world began a few thousand years ago, despite the precision apparently achieved by the bishop: he said creation occurred during the night before 23 October 4004 BCE. Remarkably, almost half of the American population still pretty much believes

Ussher's estimate, which was based on his interpretation of the genealogy of people featured in the Bible.

Thomson did not accept Darwin's theory either, possibly because of religious reasons. He therefore set out to demonstrate by mathematical calculation that Darwin couldn't possibly be right.[12] Thomson was one of the most influential and accomplished physicists of his day; he had become professor of mathematics at the tender age of twenty-two, and among his accomplishments he could count nothing less than the co-discovery (with Clausius) of the second law of thermodynamics, one of the most fundamental and enduring principles in the history of physics (and, ironically, one still misused by creationists today to deny evolution). He was someone Darwin simply had to reckon with.

Thomson's calculation of the age of the earth was based on what was then considered a reasonable assumption: that the sun was burning chemical fuel, that is, that it behaved like a giant candle. Having a good estimate of the size and mass of the star, Thomson concluded that it couldn't last more than 100 million years, far too short for anything like the Darwinian process to produce the variety of living organisms known from the fossil record and from contemporary natural history. Darwin, like most biologists since, was not good at math (his masterpiece, *The Origin of Species*, doesn't contain a single equation). Therefore he asked his son George, who was much more mathematically inclined, to check Thomson's calculations. Unfortunately, the verdict was disheartening: Thomson's procedure was correct.

Still, Darwin did not abandon his theory (he had much too much empirical evidence in its favor), evidently assuming that something *must* have been wrong with Thomson's calculation. It turns out that Darwin was correct: as we now know, the energy that keeps stars burning is not chemical, but nuclear, generated by the fusion of atoms of hydrogen and other substances (depending on the type and age of the star). Ironically, when radioactivity was discovered at the end of the nineteenth century, it was George Darwin—by then a professor of astronomy at Cambridge—who published an article in the prestigious *Nature* magazine vindicating his father's skepticism of Thomson's conclusions.[13]

The point of the story is not simply that Thomson was wrong and Darwin right—psychologically satisfying to a biologist like me as this may be. Rather, what is important to note is that Thomson committed a blunder because he rejected an overwhelmingly corroborated theory based on a dubious assumption (and perhaps on his ideological preconceptions).

There was much more evidence in favor of Darwin's theory of evolution than of Thomson's theory of star fuel. Moreover, for all his kindness, Thomson refused to admit that he was wrong even after Darwin's son published his rebuttal in *Nature*, thereby confirming the constructivist point that individual scientists are not necessarily any more objective than the guy next door. We shall see later on how science itself can still claim a high degree of quasi-objectivity, despite the fact that its practitioners are not objective machines, but instead are emotionally and subjectively after the same three universal rewards sought by humankind: fame, money (or material resources), and sex (not necessarily in that order).

The history of evolutionary biology offers another instructive story of blunder and embarrassment, this one complete with a clever hoax and a culprit who has never been found: the story of the so-called Piltdown Man. Whenever one debates creationists (admittedly a questionable, yet hard-to-kick habit), one is bound to run against the infamous Piltdown forgery. This is the case of an alleged missing link between humans and so-called lower primates, which was found in England (near Piltdown, in fact) and announced to the world on 18 December 1912. The announcement was made by Arthur Smith Woodward, a paleontologist at the British Museum of Natural History, and Charles Dawson, the local amateur paleontologist who had actually discovered the fossils. The problem is—as creationists never tire to point out—that the "Dawn Man of Piltdown" (scientific name *Eoanthropus dawsoni*, in honor of its discoverer) turned out to be a fake. Moreover, it took scientists four decades to figure this out, an alleged example of what happens when one takes science on faith, in this case the theory of evolution. But is this view accurate?

Many scientists are rather embarrassed by the Piltdown debacle, apparently feeling guilt by association, considering themselves indirectly responsible for whatever goes wrong in their chosen profession. And yet, Piltdown should instead be presented in all introductory biology textbooks as a perfect example of how science actually works, as we shall see in a moment.[14] First off, we need to realize that before Piltdown, very little was known of the human fossil record. When Darwin wrote *The Descent of Man* he had to rely largely on comparative data with other living species of primates, for then only the clearly almost-human Neanderthals were known to paleontologists. A few years before Piltdown, two important discoveries were made: that of "Java man" in 1891 and that of "Heidelberg man" in 1907, but neither of these was very ancient. When a significantly older set of prehuman remains was allegedly found at Piltdown, the scientific world was simply ready for the discovery. It was what practitioners in the field had expected, something that surely the perpetrator of the hoax knew very well and magisterially exploited.

The Pitldown findings occurred on more than one occasion and at two separate sites, yielding fragments of skulls, of a lower jaw, and even of stone tools associated with the "culture" of the predawn men. While there were skeptics from the beginning, the hoax was simply too elaborate and cunningly put together to raise the suspicions of a significant number of paleontologists—at least initially. National pride probably also played a role in a professional establishment that at the time was dominated by British scientists, with the British Museum being the epicenter of all the activities surrounding the study of the Piltdown fossils. Understandably, the British scientists involved were only too happy to claim their motherland as the place of origin of all humanity.

Yet, suspicions about the genuineness of *Eoanthropus dawsonii* grew until a group of researchers including Wilfrid Le Gros Clark, Kenneth Oakley, and Joe Weiner applied stringent chemical tests to the remains, demonstrating that the "fossils" had been planted and chemically altered to make them seem appropriately ancient: the Dawn Man was nothing but a perfectly ordinary human skull paired up with a somewhat unusually small jaw from an orangutan. What Weiner and colleagues couldn't say for sure was who carried out the hoax, although a strong case was made by Weiner that the perpetrator was none other than Dawson himself.

Be that as it may, what does this story tell us about how science works? On the negative side, it is painfully clear that science depends on an assumption of honesty on the part of its practitioners. Peer review is designed to uncover methodological or reasoning errors, not possible frauds. But since science is, once again, a human activity, egos, financial reward, and the search for glory—however brief—are still to be reckoned with. As Piltdown and other forgeries have shown, scientists are continuously open to the possibility of someone fooling them by not playing the game by the rules.

On the other hand, science is a social activity unlike any other that human beings engage in: it is a game of discovery played against a powerful but neutral opponent—nature itself. And nature cannot be ignored, at least not for long. The reason suspicions kept mounting about the true origin of the Piltdown remains was that the more paleontologists uncovered about human evolution, the less Dawn Man seem to fit with the rest of the puzzle. In a sense, the very factor that made the acceptance of

Eoanthropus dawsonii so fast in the beginning—because it seemed to be the much sought after "missing link" in human evolution—was also the reason why, four decades later, scientists kept pursuing the possibility that it was not genuine after all. While forty years of delay may seem an inordinate amount of time, they are but the blink of an eye when compared to the history of the human pursuit of knowledge. Moreover, it is important to note that it was scientists who uncovered the hoax, not creationists, which is both an immense credit to the self-correcting nature of science and yet another indication that creationism is only a religious doctrine with no power of discovery.

This is, then, why Piltdown—far from being an embarrassment to the scientific community—should be prominently featured in biology textbooks: it is an example of how the nature of science is not that of a steady, linear progression toward the Truth, but rather a tortuous road, often characterized by dead ends and U-turns, and yet ultimately inching toward a better, if tentative, understanding of the natural world.

The problem is that some of those U-turns can be painful to society when the consequences of scientific blunders don't stay neatly confined within the rarefied atmosphere of the ivory tower, but spill into the everyday social world with sometimes disastrous consequences for human welfare. The quintessential case study of this dark side of science is the eugenic movement that swept America during the first half of the twentieth century. Eugenics is often referred to as a "movement" (rather than a science) for a good reason: it was weak on science, but strong in the area of public outreach and even political intervention. It was a political ideology cloaked in the shining mantel of science, a disguise that all pseudoscience attempts to don, from the anti-HIV movement to the intelligent design movement.

From Scientific Blunders to the Immoral Use of Science: Eugenics, Past and Future

The history of eugenics is complex, as it is a uniquely fascinating blend of science, pseudoscience, politics, social factors, and historical contingencies. The reason it is relevant to our discourse is not just that it provides a perfect example of the social perniciousness of scientific blunders, but that many of the same issues confront us again as part of an ongoing discussion on whether and to what extent the modern science of genetic engineering should be allowed to modify the germ line of humanity. Modern technology could potentially achieve in a much more direct and scientifically sound way what eugenicists wished for but could not realize because of flawed science and historical accident. It is incumbent on all of us to understand the science in order to make a more informed ethical choice for our future.

The intuitive idea behind eugenics had already occurred to Darwin, at least according to the co-discoverer of natural selection, Alfred Russel Wallace, who wrote in an article published in 1890: "In one of my latest conversations with Darwin he expressed himself very gloomily on the future of humanity, on the ground that in our modern civilization natural selection had no play, and the fittest did not survive. Those who succeeded in the race for wealth are by no means the best or the most intelligent, and it is notorious that our population is more largely renewed in each generation from the lower than from the middle and upper classes."[15] It is hard to separate Darwin's from Wallace's own opinions here, but it is certainly the case that one of Darwin's sons, Leonard, dedicated his 1926 book, *The Need for Eugenic Reform*, to his father (who was long dead at that time). Interestingly, the recognized father of eugenics and the man who coined the term in his 1869 *Hereditary Genius*, was Darwin's cousin, Francis Galton.

Eugenics was founded on the basis of two sound scientific fields: evolutionary biology, rooted in Darwin's theory of natural selection; and the principle of heredity foundational to the emerging discipline of genetics, especially following the rediscovery of Mendel's work (after the publication of Galton's book) on the transmission of traits in successive generations of pea plants.[16] There was a third emerging scientific discipline that contributed to form the conceptual backbone of eugenics: statistics. Galton's own research, as well as that of Galton's most famous and influential student, Karl Pearson, were innovative because they were based on quantitative descriptions of variability in human populations, which helped put the whole field of eugenics (and, for that matter, the legitimate one of genetics) on a respectable mathematical footing. Unlike evolutionary biology, genetics, and statistics, however, eugenics differentiated itself from the very beginning because it was more about political advocacy than novel scientific research. Consider this declaration from the 1937 conference of "publicists" of the American Eugenics Society: "Man has been given a new responsibility for controlling human evolution, a responsibility he did not expect and which he is as yet wholly unprepared to discharge. Eugenics desires to prepare public opinion for the discharge of this new responsibility."[17]

Indeed, a good case can be made that eugenics had the impact that it did, in particular in the United States, because the sociopolitical situation was favorable to it, and that it ended with World War II also due to political factors, not because of any change in the science. The eugenic movement reached its peak in the USA during the 1920s, when its ideas got the ear of politicians at the local and national level (including, for example, presidents Coolidge and Hoover), as well as the financial support of magnates like George Eastman (founder of the Kodak company) and John D. Rockefeller Jr. The popularity of eugenics was due in no small measure to the fact that Americans were embracing the naïve scientific optimism born of the new discoveries of the late nineteenth and early twentieth centuries, coupled with a sense that applied science—chiefly in the form of engineering—was about to improve the human condition in spectacular ways, like the building of transcontinental railways and the opening of the Panama Canal. Tellingly, eugenics—which is now associated with regressive politics and racial prejudice—was actually embraced by social progressives to ameliorate the huge social problems posed by wave after wave of immigration, not to mention the economic and social disaster that was the Great Depression. For instance, the intellectual and progressive magazine the *Nation* invited prominent eugenicist Herbert Jennings to write an article for their "What I Believe" series, meant to showcase what highly respectable intellectuals thought about issues of the day.

Yet, to the modern ear eugenicists sound anything but progressive. For instance, a document circulated by the movement in Nassau County, New York, read in part: "castration of the male and ovariotomy will [eventually] be accepted as within the province of the state and these operations would, at once, tremendously diminish the amount of crime."[18] An article by Charles Davenport, director of the Bureau of Eugenics at Cold Spring Harbor (also in New York), funded with money from the Carnegie Institution, stated that "a clear cut presentation of the facts as to the results of reproduction of imbeciles [should] be placed in the hands of each state legislature to the end that at least female imbeciles in general be prevented from reproduction either by restraint during the reproduction period or by sterilization."[19] The misogynistic emphasis on female rather than male "imbeciles" is interesting, but the basic idea was that "illicit" unions with "feeble-minded" individuals could produce generations of undesirables to burden society.

Eugenicists weren't just about propaganda, they catalyzed significant political action. As early as 1896 the state of Connecticut passed a eugenic law according to which anyone who was epileptic, imbecile, or feeble-minded could not obtain a marriage license. Indiana passed a law in 1907 (the first of thirty-three states to eventually do so) that enacted forced sterilization of people identified as genetically inferior. In 1927 the U.S. Supreme Court upheld a similar law passed in Virginia (the *Buck v. Bell* case), and we have to wait until as late as 1967 (*Loving v. Virginia*) for the highest court in the nation finally to declare eugenic laws unconstitutional. By then, more than 60,000 people had been forcefully sterilized in the United States.[20] This compares with 350,000 sterilizations under the Nazi in Germany before World War II, a fact that did not escape some defendants at the Nuremberg trials, who claimed to have been inspired by the American example (if so, of course, the Nazi perfected the "example" beyond belief, causing immeasurable suffering and death during the Holocaust).

American eugenicists were so politically effective because they ran a savvy public campaign from the very inception of the movement. The Popular Education Committee of the American Eugenics Society recommended in 1930 a national campaign of influence in the media, targeting the two thousand circulating American papers, particularly the top three hundred or so, with a combined readership of twenty-four million people.[21] The committee laid out a detailed strategy, including monitoring progress by using a clipping service (this was in pre-Google times) and encouraging the publication of highly visible articles authored by prestigious figures favorable to eugenics in various magazines. This is not too different from the so-called Wedge Strategy that the Discovery Institute of Seattle, Washington, has been attempting to implement in recent years to promote the idea of intelligent design as an alternative to evolutionary theory.[22] As in the case of eugenics, the ID movement has little (in fact, nothing) to do with science and much to do with ideological propaganda, in this case religiously motivated.

And speaking of religion, eugenicists didn't fail to exploit the pulpit in the service of their cause either. In what turned out to be another spectacularly successful campaign, the American Eugenics Society launched a "sermon contest" to enlist the country's clergy and spread the gospel of eugenics. The society received more than five hundred entries, and it is instructive to quote from some of them:[23]

> "The aim of eugenics and the aim of the Christian Church are one in that both seek to bring the Kingdom of God down to earth and to people it with an abler and happier and more wholesome humanity."

> "The Bible is a book of eugenics . . . Christ was born of a family that represented a long process of religious and moral selection."

> "The religion of Jesus is concerned more with the nature of man than with its nurture. And so is eugenics."

And so on and so forth. If this sounds like nonsense, it is not very different in tone and intellectual content (such as it is) from what so many contemporary preachers are telling the American people about the "lie" of evolution and the truth of "creation science." Interestingly, when the eugenicists saw their influence decline, they attempted to reinvent themselves by avoiding the term "eugenics" and using instead more palatable buzzwords like "human environment," or by pushing their (disguised) ideas directly to the public, for instance by writing about when to avoid marriage in publications like *Good Housekeeping*. Analogously, the creationist movement has constantly been reinventing itself throughout the twentieth century, adapting its terminology after every major legal defeat[24] and addressing the public by a variety of direct and indirect tactics. Of course, unlike eugenics, creationism just doesn't seem to lose popular support, at least in the United States.

Eugenics eventually met its demise largely, it can be argued, because of the atrocities of Nazism before and during World War II. As Edward Caudill points out in *Darwinian Myths*,[25] the movement faced a number of unpalatable choices from which they ultimately could not extricate themselves: on the one hand, they could not criticize Hitler too vocally, under the penalty of undermining their own arguments in favor of a science-led improvement of the human race; but of course an open endorsement of Nazi Germany during WWII would have been a public relations suicide. At the same time, they failed to draw the distinction between the Nazi program of extermination of entire inferior *races* and the more "moderate" eugenicist project of improvement *within* a particular race. That distinction proved to be too subtle for a public that increasingly identified eugenics with Nazism.

Interestingly, the Nuremberg trials did not spell the end of eugenics. The idea of a fundamental biological intervention to alter the future of the human race has recently come back to the forefront of public discourse because of the stunning discoveries of molecular genetics. Since James Watson, Francis Crick, and Rosalind Franklin discovered the structure of DNA, the famous double helix, in 1953, molecular biology has arguably provided the most spectacular advances of late-twentieth-century science and equally likely the advances that will matter the most to human welfare in the near future.

The promises of human genetic engineering have so far been dramatically exaggerated. Although we know how to splice foreign DNA into a cell, we are a long way from being able to effectively and reliably fix human genetic diseases by repairing mutant genes or replacing them with functional parts off a shelve, as if we were in an auto mechanic's shop. We do have complete sequences of human genomes, and more will be produced ever faster and more cheaply during the forthcoming years, but that doesn't even begin to tell us how human beings develop from a fertilized zygote into a functional adult with billions of cells of a variety of types and functionalities. And without a much better grasp of developmental as well as molecular biology, any attempt to mess around with the human genetic line is likely to be disastrous on technical grounds, not to mention the obvious ethical implications.

Still, the sense is that our species is on the verge of a fundamental revolution in evolution: for the first time in Earth's history, living organisms will be able not just to change their environment as it best suits them (ants and beavers have been doing that for a long time, to mention just a couple of obvious examples), but to actually alter their hereditary information and, quite literally, take evolution into their own hands. This is precisely what eugenicists wished to do, but the modern science of molecular genetics will make it possible in ways that are much more refined—and potentially effective—than any follower of Francis Galton ever imagined.

The relevant point here is that Galton-style eugenics, if implemented on a wide scale, would have been an incredibly inefficient approach to the "betterment" of the human race (whatever that might mean), and probably would not have worked at all. This is because the science behind the original eugenics was rather shaky. The model was essentially that of plant and animal breeding as it has been practiced by humanity since the dawn of civilization and the invention of agriculture. Darwin himself introduced the new concept of natural selection in direct analogy with artificial selection, and the parallel does work because of the similarities between the two processes (except, of course, that natural selection is a *natural* process, not guided by a conscious intelligence). The above-mentioned confluence of evolutionary theory, genetics, and statistics provided eugenicists with a sound scientific basis for their proposals—which is, of course, an entirely different issue from that of the morality of the eugenic project. Where they went wrong, scientifically, is in their assumption that complex human behavioral traits would turn out to have the same sort of relatively simple genetic basis that Mendel's pea traits have and that one could therefore breed human beings pretty much in the same way as cattle or horses. In reality, human behavior is shaped by highly complex interactions between genetics and environment, the so-called nature-nurture problem. Moreover, the conditions eugenicists focused on ("feeble-mindedness") were both poorly defined, and hence difficult to meaningfully quantify, and more likely the result of poverty and lack of education than of "bad genes."

These technical objections to the soundness of the eugenic program carry much less force when we come to the potential use of genetic engineering to produce "Homo 2.0." In this case the parallel is not with the long and messy process of artificial breeding, but with the fast, surgical, and highly successful research program that has brought us (for good and for ill) genetically modified organisms (GMOs). Indeed, the ongoing debate about GMOs is a good model for what, on a much more dramatic scale, might soon be the debate over the genetically modified human, GMH. The lines of the debate in the case of GMOs are familiar: on the one hand are several (but by all means not all) scientists and the industry, assuring the public that there is nothing to fear from the new technology and that the skeptics are simply irrational, fear-mongering Luddites, scared of progress and science. On the other side we have said Luddites, who don't trust either science or industry and who see a return to natural foods as the only hope for humanity.

A sensible intermediate position is possible in the case of GMOs, though it remains to be seen if this is true with regard to GMHs. Every new technology does pose risks, sometimes unknown ones (until they occur), and to deny it is either naïve (possibly the case of most pro-GMO scientists) or disingenuous (arguably the situation for the pro-GMO industry). Moreover, there are good, rational reasons to maintain at least a moderate level of distrust of large biotechnology firms, given both their past record and their obvious profit motives. However, a return to nature for nature's sake would deny most of what makes modern life pleasant and safe (think of vaccines, artificial means of transportation, computers—all highly unnatural); moreover, the equation of "natural" with "good" is an instance of the same naturalistic fallacy that we have encountered before, and it is easy to disprove: there are plenty of living organisms (from viruses to poisonous mushrooms) that are as natural as they could be and of which nonetheless it is wise to steer clear.

Similarly, when it comes to the possibility of Homo 2.0, the questions to address are not just scientific, but ethical. As a scientist, I do not see any technical reason why the production of GMHs is not possible, or even within grasp of modern science. As a citizen and philosopher, I do see plenty of good reasons to have a conversation about the ethics of human genetic engineering. The following are examples of what such a conversation should *not* look like. Here is author Lee Silver on a possible future scenario involving two new classes of citizens, which he calls "GenRich" and "Naturals," for reasons that do not need further elaboration:

> The GenRich—who account for 10 percent of the American population—all carry synthetic genes . . . that were created in the laboratory . . . All aspects of the economy, the media, the entertainment industry, and the knowledge industry are controlled by members of the GenRich class . . . Naturals work as low-paid service providers or as laborers, and their children go to public schools . . . If the accumulation of genetic knowledge and advances in genetic enhancement technology continue . . . the GenRich class and the Natural class will become . . . entirely separate species with no ability to cross-breed, and with as much romantic interest in each other as a current humans would have for a chimpanzee.[26]

If this sounds like a sci-fi, nightmarish scenario straight out of H. G. Wells's *The Time Machine*, read on. Silver is serious: "Anyone who accepts the right of affluent parents to provide their children with an expensive private school education cannot use 'unfairness' as a reason for rejecting the use of reprogenetic [genetic augmentation] technologies. Indeed, in a society that values individual freedom above all else, it is hard to find any legitimate basis for restricting the use of reprogenetics."[27] Actually, it is very easy to find plenty of legitimate ethical reasons why a society would want either to ban "reprogenetics" or tightly regulate it and make it available to every one of its citizens. Silver is making the common mistake of equating unbridled capitalism with a free society (which, incidentally, is not a society in which one values individual freedom "above all else").

I am not suggesting that human genetic engineering is intrinsically immoral any more than I would suggest that using any other life-enhancing technology or medical advancement is immoral. If anything, the real problem, as is often the case, is the difficulty of guaranteeing access to all. However, there surely are difficult issues hinging, for instance, on the distinction among at least three possible uses of human genetic engineering: the elimination of hereditary diseases, "vanity" selection, and actual improvement of the species. It is hard to argue sensibly against the first use of such technologies: who would actually want, if given the choice, to have a child affected by, for instance, Canavan disease? This is an abnormality of the nervous system caused by a defective enzyme that fails to break down a substance known as N-acetyl aspartate. The accumulation of the substance in the brain interferes with the growth of myelin, an insulator that facilitates the propagation of electrical stimuli in the central nervous system. As a result, children affected by Canavan disease quickly become mentally retarded, lose motor skills, and become subject to paralysis and seizures. It so happens that we already know how to prevent Canavan disease: a simple genetic screening of the parents can establish if they carry the defective gene on chromosome 17. If both of them do, there is a 25 percent chance (in accordance with Mendel's laws) that their child will be affected, so all the couple needs to do to prevent the disease is not to have children (or adopt, or look for another partner). Still, a genetic engineering approach to the problem would allow anyone to have children with anyone else, regardless of whether they are carriers of the defective gene that causes Canavan disease, because that copy of the gene could be repaired or substituted in the fertilized zygote. Objecting to such procedure on moral grounds would be similar to objecting to vaccination on the ground that God wants us to suffer from the diseases He invented (the absurdity of which has not stopped people from actually defending such "reasoning").

The situation is clearly different (again, ethically, not scientifically), when it comes to vanity uses of human genetic engineering. Consider, for instance, a couple who wishes to have a daughter with blue eyes, even though there is no chance this will happen naturally, given the genetic constitution of the two parents. On the one hand, it is difficult to make the case that such intervention would in any way be harmful to the parties involved. On the other hand, it seems like a waste of resources (genetic engineering is not going to be a cheap, off-the-shelf technology any time soon) and may be objected to on the grounds that it commercializes humanness, something we already have enough of as it is.

Finally, contemplate the further logical step: if we can soon learn to eliminate genetic defects, the same techniques would presumably allow us to "improve" the natural genetic constitution of a human being. One could quibble on what exactly "improvement" means in this context, but again it would be hard to see cogent objections against, say, making people naturally capable of maintaining healthy levels of cholesterol, or equipping them with sturdier bones or longer-lasting hearts, and so on. This doesn't necessarily mean that we would lose genetic variation in the human population (indeed, we could artificially *increase* the number of human variant traits), nor, of course, that we would be building armies of easily controllable zombies (nobody has discovered the "free will" gene—yet). But the very possibility of us meddling with Homo 1.0, the outcome of millions of years of evolution, should at least give us pause. As a species, we do not have an exactly reassuring track record when it comes to the implementation of some of our far-reaching technological advances (think of nuclear weapons or the industrial causes of global warming), so it stands to reason that we should be cautious when we come to nothing less than the possibility of taking control of our own long-term biological future.

Throughout this chapter we have seen that critics of science do have several good points. Science indubitably is a human activity, hampered by epistemic limitations and influenced by the ideologies and personal inclinations of those who practice it. Scientific research is not a steady march toward Truth, and indeed it is difficult to understand how it is even possible to measure scientific progress if one thinks of truth as a correspondence between our ideas and the reality "out there." Scientists are prone to blunder because of limited information or sheer oversized ego, which encourages them to make confident statements about matters concerning which they would be better served by caution or silence. Most important, science has dramatic effects on social welfare, and by no means are all of those effects benign, nor should they be under the exclusive or even primary control of scientists themselves (who, after all, are citizens like anyone else, with no special intrinsic privileges). Criticism of science is, therefore, a crucial component of an open society, and it arguably helps science itself both as a process of discovery and as one of human betterment. All of this said and done, however, we should not throw the baby out with the bathwater. Science remains by far the most effective way of gaining knowledge (and power, as Francis Bacon famously pointed out) over the natural world and improving the human condition. That is why it is important to rein in the extreme versions of science criticism that have developed during the past couple of decades and that I have referred to with the generic term of (social) constructivism. To that task we turn in the next chapter.

CHAPTER ELEVEN

The Science Wars II

Do We Trust Science Too Little?

The only absolute truth is that there are no absolute truths.—Paul Feyerabend, philosopher

When one analyzes [postmodernist] writings [on science], one often finds radical-sounding assertions whose meaning is ambiguous and that can be given two alternative readings: one as interesting, radical, and grossly false; the other as boring and trivially true.—Alan Sokal, physicist

In the previous chapter we saw that there are, indeed, good reasons to be skeptical of grandiose claims about the power of science. Even highly respected scientists like Lord Kelvin are sometimes prone to make absolute declarations of truth that are then spectacularly falsified by new discoveries. Scientists and nonscientists alike can be fooled by hoaxes such as the Piltdown Man for decades at a time, while the grim history of eugenics should be a permanent warning about the costs in terms of human suffering that can result when ideology takes over science. We have seen that science does not afford a "God's-eye view" of the world and that scientists themselves are not always successful in being detached, rational agents interested only in the pursuit of truth.

Nonetheless, there is much value in what science does and how it does it, and there is much nonsense that goes under the general label of "science criticism." In this second chapter on the science wars, I will briefly examine some of the most outrageous statements put forth by science critics to see just how far the postmodern assault on science can go. Then it will be time to take stock of the pros and cons of this debate and begin to work our way toward a balanced conception of science in which rational criticism of what scientists do is an integral part of the democratic discourse in an open society and in which science itself is reconceived as neither inhumanly objective nor hopelessly doomed by human subjectivity. Science can instead be thought of as offering a perspective, the best perspective so far available to humanity, on how the world really works.

Postmodernism: How Bad Is It, Really?

The now classic way to introduce the public to the excesses of postmodernist critiques of science is through a recounting of the so-called Sokal affair. Alan D. Sokal is a physicist who, in the mid-1990s, got tired of all the nonsense he kept hearing about science from some vocal academics in various humanities departments. He decided to strike back with a bit of humor, probably without realizing that his name would rapidly become an icon (for good or ill) in the science wars. Sokal wrote a piece entitled "Toward a Transformative Hermeneutics of Quantum Gravity"[1] and submitted it to the prestigious postmodern journal *Social Text*. "Hermeneutics" means the study of texts, and in modern philosophical parlance it refers to an approach in which not only texts, but also artifacts (like works of art) and theories (including scientific theories) are to be understood as products of a given cultural and historical set of circumstances. Quantum gravity is a field in theoretical physics that aims at unifying quantum mechanics, a theory concerning three of the four fundamental forces of nature (the strong and weak nuclear forces, plus electromagnetism), and general relativity, which describes the fourth force, gravity. But what could one mean by applying hermeneutics to quantum gravity?

As it turns out, nothing at all; Sokal had written—on purpose—nothing more than thirty-five pages of nonsense and non sequiturs. The editors of *Social Text* were lured into accepting the article, however, simply because it was written by that rarest of academic breeds: a hardcore scientist who dares to criticize his own discipline from a philosophical perspective. The editors gleefully swallowed the bait, only to be crushed when Sokal exposed the hoax in the magazine *Lingua Franca*. Thus the Sokal affair was immediately catapulted into the limelight, providing an endless source of amusement for scientists and of rage for postmodernists. Sokal himself, however, has a sober assessment of his own stunt, pointing out that "some of my overenthusiastic supporters have claimed too much"[2] and adding that "it doesn't prove that the whole field of cultural studies . . . is nonsense." Indeed, just in the same way that the Piltdown hoax doesn't prove that evolutionary biology is nonsense, pace the creationists. According to Sokal, all his little experiment did was prove that "the editors of one rather marginal journal were derelict in their intellectual duty, by publishing an article . . . they could not understand . . . solely because it came from a 'conveniently credentialed ally' [as *Social Text* coeditor Bruce Robbins later candidly admitted], flattered the editors' ideological preconceptions and attacked their 'enemies.'"

Here Sokal may have been a bit too magnanimous, however. To begin with, *Social Text* is far from being just a "marginal" journal in the postmodernism field. Most importantly, though, Sokal's paper was largely developed by cobbling together actual quotes from published postmodernist authors. It is this detail that is damning for the whole field, because it exposes the fact that many (though by no means all) "science critics" actually know little or nothing about science, yet get away with making outrageous statements because of the preconceived ideological commitment of their readers. Of course, it is easy to play the "out of context" game by willfully misquoting authors and making them sound as if they were asserting sheer nonsense or taking positions they set out to criticize. But in the case of some prominent postmodernist writers, it is hard to imagine in what possible context some of their utterances would make any sense at all. A few selected gems from the vast collection available will make my point, just as they helped Sokal.[3]

Harry Collins, a sociologist who coauthored the insightful *The Golem: What You Should Know about Science* (a book about the limits and human aspects of the scientific enterprise), wrote, as we saw in the last chapter, that he is sympathetic to the relativist school of sociology of science, which maintains that "the natural world has a small or non-existent role in the construction of scientific knowledge."[4] Now imagine if one were to take that statement seriously when assessing, say, the difference between the Ptolemaic and Copernican systems in astronomy: could it really be that the *fact* that the earth rotates around the sun, and not the other way around, played "a small or non-existent role" in the acceptance and endurance of Copernicus's theory?

Bruno Latour, a French sociologist, is one of the most egregious offenders in the science wars.[5] He made a fool of himself by commenting on Einstein's theory of relativity without displaying any understanding of what the theory is actually about. Einstein proposed (and plenty of experimental evidence has verified) that there is no privileged frame of reference in the universe, that is, that all measures of time and space depend on the motion of the observer. Thus, unlike in the case of Newtonian mechanics, neither time nor space are absolute quantities (hence the word "relativity" that characterizes Einstein's theory). The theory has a number of counterintuitive consequences, including the famous "twin paradox," according to which a twin traveling on a spaceship near the speed of light would age much more slowly than a second twin left on Earth. This is because within their two frames of reference (the spaceship and planet Earth), time would progress at different paces, literally slowing down with higher and higher speeds.

Now here is a comment, quoted by Sokal, that Latour made on Einstein's theory of relativity, in which he takes to task "[Einstein's] obsession with transporting *in*formation through *trans*formation without *de*formation; his passion for the precise superimposition of readings; his panic at the idea that observers sent away might betray, might retain privileges, and send reports that could not be used to expand our knowledge; his desire to discipline the delegated observers and to turn them into dependent pieces of apparatus that do nothing but watch the coincidence of hands and notches."[6] This would be comical if it weren't such a sad example of the misunderstanding of science by a nonscientist. First of all, Latour is criticizing a popular book by Einstein, not a technical article, presumably because Latour simply cannot understand a technical paper in fundamental physics (full disclosure: nor can I, nor most people living on the planet—but some of us try not to write commentaries about things we do not understand). More importantly, Einstein was explaining his theory through a *thought experiment*, using hypothetical human observers to help describe the nuances of the concept. Obviously observers need not actually be there for the theory to work. Latour's talk of "panic," "betrayal," and "privileges" is complete emotive nonsense. This is physics, not a Shakespearian tragedy.[7]

Some feminist philosophers who embrace the postmodernist approach have also done a good deal of damage in the context of the science wars. The field is ripe with acrimony and controversy, with feminist scholars such as Helen Longino (more on her contributions below) saying very interesting things about the nature of science, and others making little more sense than the above-quoted Latour. Janet Biehl, a social ecologist and a feminist herself, rightly criticizes some of her postmodern colleagues because "Ecofeminism's sweeping but highly confused cosmology introduces magic, goddesses, witchcraft, privileged quasi-biological traits, irrationalities, neolithic atavisms, and mysticism into a movement that once tried to gain the best benefits of the Enlightenment and the most valuable features of civilization for women, on a par with thinking and humane men."[8] Then again, when feminist philosopher Sandra Harding boasts that "I doubt that in our wildest dreams we ever imagined we would have to reinvent both science and theorizing itself,"[9] one wonders where this new science is to be found and what sort of discoveries feminist scientists, as distinct from standard scientists (many of whom, of course, are both women and feminists), have made.

Harding is the same author who wrote in the *New York Times* of "the important role that sexual metaphors played in the development of modern science. They [feminist authors Carolyn Merchant and Evelyn Fox Keller] see notions of dominating mother nature by the good husband scientist. If we put it in the most blatant feminist terms used today, we'd talk about marital rape, the husband as scientist forcing nature to his wishes."[10] But perhaps it does not make much sense to "put it in the most blatant feminist terms." To begin with, as both a scientist and a man, I deeply resent the very idea that my curiosity about nature has roots in my alleged fantasies about rape and domination (and are the same fantasies haunting my female science colleagues? One can only wonder). But—for the sake of argument—let us go as far as to accept the possibility that Francis Bacon or Isaac Newton's metaphorical talk about unveiling the secrets of mother nature has some remote connection with their sexuality (for that matter, why not go Freudian all the way and suggest that they were simply longing for their mother's womb?). Does any reinterpretation of *why*, at a psychological level, scientists do what they do negate their empirical findings or theories about nature? Have feminists proposed new theories of optics or gravity to replace Newton's work? Are we building new telescopes based on innovative feminist perspectives? Asking these questions is not raising the bar too high for the very good reason that authors like Harding themselves set the bar: if they were simply claiming that there are social and historical factors, including the history of gender roles, that affect the practice of science, that would be reasonable and hardly controversial. But when one claims to be "reinventing science," one better put her money where her mouth is. So far, no such luck, and the great pronouncements of feminist science have fallen as flat as the similarly empty utterances of supporters of intelligent design.

The examples could go on and on, but no commentary on postmodernism in science can overlook the contribution, such as it is, of David Bloor, one of the promoters of the so-called strong programme in the social sciences. In an essay he coauthored with Barry Barnes they define knowledge as "any collectively accepted system of belief."[11] Notice the peculiar absence of any reference to whether such beliefs are, in fact, true or not—something that simply doesn't figure among the objectives of the strong programme. As we have seen, ever since Plato, philosophers have defined knowledge as *justified true belief*, meaning that in order to count as knowledge, one's notions have to be both true (or as likely to be true as humanly possible) and justifiable (on the basis of reason and/or empirical evidence). As Sokal points out, commenting on Barnes and Bloor, their concept of knowledge avoids any distinction between what is really out there and our understanding of it. The two certainly interact in complex ways, since all knowledge of the world has to be achieved through the cognitive and sensorial limitations of being human, which means that what I would call a "modest programme" in the sociology of science makes perfect sense and could contribute seriously to our understanding of science. But to not even incorporate any reference to the world in one's discussion of knowledge is instead an instance of the intellectual perversity that marks much postmodernist discourse.

The postmodern cultural wars of the 1990s (which to some extent are still alive today) are by no means the first time that science has been attacked on intellectual grounds (as opposed to, say, by the admittedly more worrisome stratagem of threatening scientists with being burned at the stakes). An important precedent to the postmodernist assault, and one that inspired many postmodernists, can be found in the work of a philosopher of science active during the 1960s and '70s: Paul Feyerabend.

Rebel with a Feeble Cause: The Ranting of Paul Feyerabend

Paul Feyerabend was one of the most influential philosophers of science of the twentieth century—unfortunately, I might add. While he was not a postmodernist (he reportedly did believe in an objective reality), his writings have given plenty of ammunition to post-modernists and even the few creationists who know who he was. As we shall see shortly, he bears a partial indirect responsibility for the instigation of the culture wars since he was very consciously, indeed recklessly, working to provoke a clash between science and society. As with most science criticism, his motives were probably genuine,[12] and some of his reasoning was sound and on target. But, again as is too often the case with science critics, he quickly embarked on an absurd slippery slope that ended up causing much more damage than good. This is why I'm devoting a significant amount of space to Feyerabend: he is the quintessential example of what goes wrong when science and the humanities collide.

Paul Feyerabend had an interesting personal history, which he likely would have admitted may have strongly influenced some of his intellectual positions. Born in Vienna in 1924, he was drafted into the German army during World War II, during which he was decorated with an Iron Cross. Toward the end of the conflict, he was hit by three bullets while directing traffic, and one of the bullets hit his spine. This caused him constant pain for the rest of his life and forced him to walk with a cane. It is easy to imagine that this sort of life experience would make someone angry, and anger and contempt are what Feyerabend's writings most clearly evoke (though, of course, it is hard to establish a direct causal link between one's biography and his intellectual positions). Unfortunately, neither the philosopher's anger nor his contempt amount to a good reason to buy into his most egregious intellectual positions.

Let me start by giving Feyerabend his due in terms of his thinking about science, as presented in his most influential book, aptly entitled *Against Method*, published in 1975. Feyerabend's basic idea was that there is no such thing as a unified, universal scientific method. In this he was reacting against the ideas of his former mentor, our old acquaintance Karl Popper, and he was, by and large, correct. We have seen that Popper was interested in the so-called demarcation problem—how to reliably distinguish science from pseudoscience. We have also seen why his particular solution, the idea of falsificationism, doesn't really work. Indeed, Popper's premise that philosophy of science could play a *prescriptive* role, that philosophers could dictate what scientists should do and how they should go about their business, was rather naïve and has since fallen into disfavor, replaced by a more *descriptive* philosophy of science, wherein philosophers try to understand, as opposed to dictate, how science works.

While Feyerabend was correct in rejecting the simplistic idea of a universal and ahistorical scientific method, he was wrong in his reasons for doing so. In *Against Method* he attempted to show through historical examples that new scientific ideas get accepted not on the basis of logic and evidence, but in spite of them, in a process in which the individual personalities and even oratory abilities of individual scientists play the crucial role. For instance, Feyerabend claims that Galileo "prevails because of his style and his clever techniques of persuasion, because he writes in Italian rather than in Latin, and because he appeals to people who are temperamentally opposed to the old ideas and the standards of learning connected with them."[13] There is a grain of truth in this, as one would expect, since science is done by scientists, and scientists are human beings, and human beings are prone to respond to psychological factors as much as to anything else. But to claim that this is most or all of the reason why Galileo eventually won the day is a gross misrepresentation of historical reality, as has been nicely shown by Alan Chalmers in his critique of Feyerabend.[14] Galileo succeeded because of the compelling logic of his science and because of the overwhelming empirical evidence he adduced in support of his theories, not because he was writing in Italian.

What is true of the history of science as we have come to reconstruct it is that scientists evolve their own methods and that methods that are prevalent in, say, physics, may not work as well in biology and vice versa. Science is perhaps best conceived as a *family* of activities that share a common base provided by reliance on empirically testable hypotheses and on rationally constructed theoretical frameworks. We saw early in this book, for instance, how ahistorical sciences like particle physics do work in a significantly different fashion from historical ones like geology, with mixed-approach sciences such as evolutionary biology (in which both historical research and experimentation are practiced) somewhere in between. Historians of science have shown that scientists such as Galileo, Newton, and Einstein do, over time, change the methods, both empirical and conceptual, used by their disciplines, introducing new ones and discarding those that are no longer working. But this dynamic view of science is not at all equivalent to Feyerabend's "anarchic" concept of "anything goes"—far from it.

To get a taste of the absurdity of Feyerabend's most controversial positions, as well as why he often triggers passionate responses such as my own, let us briefly look at an essay he wrote in 1974, before the publication of *Against Method*.[15] The essay is provocatively entitled "How to Defend Society against Science," and it's Feyerabend at his legendary best (or worst, depending on your point of view). Feyerabend begins by telling his readers (actually, his audience, since the essay started out as an invited talk) how he arrived at what he was about to present. If you were thinking of a tortuous and fascinating intellectual journey, think again. "About a year ago I was short of funds. So I accepted an invitation to contribute to a book dealing with the relation between science and religion. To make the book sell I thought I should make my contribution a provocative one and the most provocative statement one can make about the relation between science and religion is that science is a religion. Having made that statement the core of my article I discovered that lots of reasons, lots of excellent reasons, could be found for it. I enumerated the reasons, finished my article, and got paid."

Let us deconstruct this a little bit. Feyerabend is claiming that he came to the (preposterous) position that science is a religion because he wanted some money and figured that one way to sell books is to be controversial (that part, I'm afraid, is at least partially correct). He then convinced himself that there were actually pretty good (no, "excellent") reasons for his position after all. This is such a transparent example of rationalization and intellectual travesty that one would be justified in closing the book at that point and disregarding whatever else Feyerabend had to say. But that would be a mistake, because we would not then learn as much about the minds of some "science critics." So let us continue to explore his thought processes through his own words.

"Consider the role science now plays in education," Feyerabend continues.

> Scientific "facts" are taught at a very early stage and in the very same manner as religious "facts" were taught only a century ago. . . . In society at large the judgment of the scientist is received with the same reverence as the judgment of bishops and cardinals was accepted not too long ago. . . . Do not be misled by the fact that today hardly anyone gets killed for joining a scientific heresy. This has nothing to do with science. It has something to do with the general quality of our civilization. Heretics in science are still made to suffer the *most severe* [emphasis in the original] sanctions this relatively tolerant civilization has to offer.

Again, some deconstruction is in order. Apparently, for Feyerabend "facts" about mythical supernatural deities spitting on mud and thereby breathing life into humans are on par with "facts" about the Earth being a planet that rotates around a star. He seems oblivious to the reality that in our society the opinions of scientists are most certainly not accepted in the same way as the pronouncements of religious figures, as clearly demonstrated by the rise of creationism in the public arena in the United States when Feyerabend was writing this nonsense. Moreover, and most outrageously, he has the gall to compare academic debates to the Spanish Inquisition, which sounds like something straight out of a Monty Python skit and which doesn't really deserve an actual rebuttal (one should notice that Feyerabend spent most of his career at the University of California, Berkeley, a *public* institution where his position and salary were not endangered by the silliness of his ideas for a long period from 1958 to 1989, at which time he left for Europe).

Feyerabend was an optimist at heart, however. "The situation is not as hopeless as it was only a decade ago. . . . We have learned that there are phenomena such as telepathy and telekinesis which are obliterated by a scientific approach and which could be used to do research in an entirely novel way. . . . And then—is it not the case that the Church saved souls while science often does the very opposite?" No, Paul, it isn't the case at all. Besides the fact that I'm not sure of what it even *means* to say that science damns souls (or, for that matter, that the church saves them), Feyerabend's uncritical acceptance of pseudoscientific notions such as telepathy and telekinesis is a spectacular illustration of what is wrong with the "anarchic" approach to knowledge: it leads to a hodgepodge of good stuff and sheer nonsense, with no way to separate the chocolate from the manure, so to speak.

Feyerabend goes on to call for a "formal separation between science and state" analogous to the separation of church and state affirmed by the United States Constitution. Were that to actually occur, all public funding of science would cease, and science would be left in the hands of large for-profit corporations, the very same guys who swore that there is no link between smoking and lung cancer, and who had the data to "prove" it. Indeed, Feyerabend shamelessly encouraged creationism with his "three cheers to the fundamentalists in California who succeeded in having a dogmatic formulation of the theory of evolution removed from the textbooks and an account of Genesis included" (though to his credit he did also add that "I have no doubt that they [the creationists] would be just as dogmatic and close-minded if given the chance"). The reference was to a temporary victory for young earth creationists, soon reversed by the California courts, but one has to wonder why this is the sort of intellectual and educational landscape that Feyerabend promotes in order to build his intellectual utopia.

Not surprisingly, for Feyerabend, the whole concept of public education is a "myth," despite the fact mentioned above that he spent most of his career in a plush job at a public institution (this is the academic equivalent of those politicians who rail against government while running for election so that they can work for the government and enjoy the many benefits and privileges that usually come with such jobs). For Feyerabend, "the method of education often consists in the teaching of some *basic myth* [emphasis in the original]," but his new social order will have none of that because "of course, scientists will not play any predominant role in the society I envisage. They will be more than balanced by magicians, or priests, or astrologers." Yes, I just hope that for the sake of consistency we will also invite magicians, priests, and astrologers to replace other experts in our society, like car mechanics, brain surgeons, and civil engineers.

Paul Feyerabend managed to be (consciously, by his own admission) controversial by saying outrageous things that he probably didn't believe (according to his autobiography, he took medications for his back pain—he didn't go to a magician). Controversy is an important part of intellectual and social discourse, and people who have the guts to say what they think regardless of how popular their positions may be deserve our respect, even when they turn out to be wrong. But to utter outrageous statements for the professed purpose of selling books is an intellectual travesty, to say the least. Yet it is precisely in keeping with the sort of characters we have discussed so far, who foment the culture wars and end up wasting a lot of people's energy and resources. Given this background, it is now time to turn to a more somber consideration of the serious issues raised by science criticism, entertaining a few good solutions that less flamboyant but more earnest thinkers have attempted to provide.

Of Paradigm Shifts and the Progress of Science

Despite the outrageousness of his approach, Feyerabend did have a point: we still need an account (as philosophers say) of how it is that science does what scientists claim it does. According to its practitioners, science is a very peculiar type of human enterprise, characterized by the ability to make progress (we understand the world better and better) and the property of being objective (i.e., its results are taken to be independent of specific cultural attitudes and backgrounds). This is most obvious in the case of technology, the applied aspects of science. It should be clear to anyone that our means of transportation, for instance, or our ability to build bridges have gotten more and more sophisticated with the passing of time. The ancient Romans built sturdy bridges all over their empire, but it took late-nineteenth-century technology to build the Brooklyn Bridge. And it is only with twentieth-century technology that we have been able to do regular transoceanic flights, or to go to the moon. But how is scientific (and hence technological) progress possible? We began this discussion in the last chapter, where we tackled the so-called correspondence theory of truth. The rest of this chapter will draw on the works of several authors to resume that line of inquiry and show that the claims that scientists make about science are in fact correct, though not exactly for the reasons that scientists themselves would adduce.

Many people without a background in philosophy have heard of the word "paradigm." Indeed, the phrase "paradigm shift" is now frequently used in the media and in business circles to indicate a major change in the way of thinking in a particular field. The concept, however, originated with philosopher Thomas Kuhn, who in the 1960s and '70s wrote about how science works. Kuhn examined actual examples of change in scientific theories and tried to make sense of how this happens based on available historical documents. For instance, he took an in-depth look at classic cases in the history of astronomy and physics, like the superseding of the Ptolemaic theory by the ideas of Copernicus and Galileo or the replacement of Newtonian mechanics with general relativity and quantum mechanics. Kuhn concluded that science proceeds according to two fairly distinct modes of operation. Much of the time, scientists are involved in the first mode, which Kuhn called "normal science." This is research dedicated to the solution of small problems (Kuhn called them "puzzles") within the generally accepted theoretical framework of the discipline, the "paradigm." So, for instance, biologists working within the Darwinian paradigm may be interested in figuring out how a particular species of invasive plants has been so successful in a new geographic area, or a physicist working within the quantum mechanical paradigm may devote her efforts to the understanding of the properties of a specific type of subnuclear particle.

Normal science may proceed for decades, sometimes more, without any momentous change. However, it also regularly incurs what Kuhn called "anomalies," puzzles that elude solution, either because of technical limitations in the instruments available to scientists or because the paradigm does not seem to provide good theoretical tools for dealing with the anomalies. Examples of anomalies include the current difficulty for evolutionary biologists in providing a detailed account of the evolution of certain structures, such as the carapace of turtles; or the fact that physicists are unable—at the moment, anyway—to reconcile some of the predictions of general relativity with those of quantum mechanics. Scientists don't usually lose much sleep over the existence of anomalies, because every paradigm will have them: since science is continuously developing, and because all scientific conclusions are best regarded as tentative, anomalies are not only a fact of life for scientists, but are a constant incentive to come up with new ways of solving the existing puzzles.

Here is where things get interesting: from time to time, according to Kuhn's reconstruction of the history of science, the number of anomalies that won't go away becomes embarrassingly high, with some problems sticking obstinately through the fabric of the reigning paradigm. When that happens, the field begins to transition to Kuhn's second mode, a phase of "revolution" in which the old paradigm is now openly questioned by an increasing number of scientists. Eventually, a new paradigm may take shape and replace the old one, thus returning the field to a state of normal science and puzzle-solving—until the next crisis and paradigm shift occur. For instance, by the time of Copernicus (early sixteenth century), astronomers had become weary of the inability of the Ptolemaic system, with its assumption that the Earth is at the center of the universe, to accurately predict the positions of the planets. Their first strategy to deal with the problem was to work within the paradigm, inserting additional calculations until the model matched the prediction. As a result, a number of hypothetical "epicycles"—additional orbits within orbits that did not actually correspond to physical characteristics of the solar system—were included in the models. Essentially, mathematical artifices were created in order to correct the mismatch between Ptolemaic predictions and actual observations of the positions of the planets. However, the more observations that were made and the more those observations became precise with improved instruments and methods, the more obvious it was that something was off with the whole Ptolemaic paradigm, as an embarrassing number of epicycles became necessary to account for the data. It was Copernicus who eventually presented an alternative hypothesis that is close to what we accept today, and it was the work of Galileo and of Kepler that perfected the system, thereby completing the paradigm shift.[16]

The Kuhnian idea of long periods of puzzle-solving punctuated by relatively short bursts of shifting paradigms may not apply to all sciences.

For instance, it is not at all clear whether evolutionary biology has gone through any paradigm shifts since Darwin: while the theory keeps being modified and augmented in scope and number of explanatory principles (in addition to natural selection, today's biologists recognize several causal mechanisms for evolution), one would be hard pressed to fit the history of that field into Kuhn's scheme, since Darwin's original ideas are still considered largely valid. Nonetheless, the basic Kuhnian model is appealing, and it is a matter of historical research to figure out the limits of its applicability. The trouble, if any, comes from a different direction entirely, best represented by Kuhn's choice of an analogy for his ideas about the process of science: Gestalt psychology.

Gestalt psychology is a theory of perception that says, among other things, that our brains perceive holistic wholes, not atomized parts of objects. Classic examples of the Gestalt effect are drawings that can be perceived in more than one way by the observer, for instance as a young woman or an old one, or a vase versus two faces looking at each other, or a cube with its faces protruding from the page or rotated inward. Different people usually see only one interpretation of the drawing and recognize the alternative only after some effort. Once someone is able to see both versions, the brain can easily switch back and forth between them in a stunning demonstration of how we interpret what we think we simply "see." Kuhn compared the two interpretations of a Gestalt figure to two paradigms, that is, two different ways of looking at the same objective data. As in the case of the Gestalt effect, most scientists are used to one paradigm (the one within which they were trained), but some at least will be able to see the alternative one, thereby sparking a scientific revolution if they can convince enough of their colleagues that the new framework is better.

And therein lies the rub: if we take Kuhn's analogy seriously, in what sense is one paradigm "better" than the other? Neither of the two alternative interpretations of a Gestalt image is better than the other for the simple reason that they are both arbitrary understandings imposed by our brains on what in reality is a set of lines or shades on paper. There is no sense in which, say, a particular drawing typically used in Gestalt demonstrations is "better" understood as two faces looking at each other, as opposed to a black vase against a stark white background. Indeed, there *are* neither faces nor vases, just a set of black lines on white that the human mind interprets as faces or vases, given a certain cultural and personal background. The situation is not helped by the fact that Kuhn sometimes referred to paradigm shifts as akin to political revolutions or, worse yet, religious conversions. Again, this doesn't sound like a model of how science makes progress, but rather fits well with the extreme postmodernist view that science is an arbitrary human activity—precisely like politics or religion. This is why Kuhn, to his chagrin, began early on to be favorably regarded by some science critics—to the point that he felt it necessary to write a postscript to the second edition of *The Structure of Scientific Revolutions* in which he said: "later scientific theories are better than earlier ones for solving puzzles in the often quite different environments to which they are applied. That is not a relativist's position, and it displays the sense in which I am a convinced believer in scientific progress."[17]

So which one is it? Was Kuhn a relativist or did he accept the idea that science does make progress? According to Alan Chalmers, in his wonderful *What Is This Thing Called Science?* it is actually both. Chalmers suggests that Kuhn was using the word "knowledge" in two distinct ways and that the resulting confusion accounts for the two apparently incompatible strands in Kuhn's thinking about science. It is worth examining Chalmers's point, not just because it illuminates the problem with Kuhn's ambiguity, but more importantly because the same ambivalence concerning what one is talking about when dealing with claims of "knowledge" is one of the root causes of the whole postmodernist mess.

The distinction Chalmers is drawing is between objective and subjective knowledge: if I say "I'm thinking of chocolate ice cream," that claim may be true, but nobody has any way to verify it because that knowledge is in my head only (at least until neurobiologists figure out a way to scan my brain and read what I'm thinking based on which neural pathways I'm using). I *know* what I'm thinking, but such knowledge is obviously subjective. On the other hand, if I say "The Brooklyn Bridge connects downtown Brooklyn with lower Manhattan," I am also making a knowledge claim, but this time it is objective knowledge in the sense that anyone can pick up a map of New York City, look at a satellite picture, or, better, yet, go there and actually verify that what I claim is in fact true.

Chalmers suggests that Kuhn is talking about objective knowledge when he says that a new paradigm represents true progress in science, as there are publicly available data that anyone can use to verify factual claims. Puzzles previously unsolved under the old paradigm will now find a solution, while puzzles that were already solved under the old paradigm should remain that way. If this latter statement did not hold, then one would simply be trading old problems for new ones, which wouldn't represent much progress.

However, when Kuhn talks about Gestalt switches and political revolutions, he is referring to the subjective knowledge that any particular scientist has of her field and its paradigm(s). So, for example, an individual scientist may not embrace a new paradigm not because there aren't objectively good reasons to do so, but because of personal, psychological motivations. Moreover, individual scientists are likely not aware of *all* the features of a paradigm, even though those features are objectively available to anyone working in the field. That accounts for the frequent case in the history of science in which scientists "discover" new implications of a theory that had been proposed decades earlier, as in the new interpretation of the photoelectric effect that led Einstein to his theory of relativity.[18] It's not that such implications were not present from the beginning (objective knowledge), it's that it took people some time to figure them out (subjective knowledge). Chalmers's conclusion is that we should disregard the (few) parts of Kuhn's theory that refer to subjective knowledge, after which we are still left with a substantial theory of how science works *and* makes progress.

If we do accept Chalmers's modification of Kuhn's theories, we also have a pretty good answer to one of the most damning issues emerging from a careful consideration of the history of science: as it turns out, *every* scientific theory ever proposed in the past has eventually been proven wrong and has given way to new theories. If that is the case (and it is), then what makes scientists think that they got it right this time around? The answer is that they don't think that at all, at least not when they reflect on the nature of scientific theories. There is no contradiction between acknowledging that science continues to discard its theories and the claim that it nonetheless makes progress, and Kuhn's ideas are in line with both claims. Scientific knowledge (in the objective sense discussed above) is provisional, and each successive theory, each newly established conceptual framework (or paradigm) represents the best understanding of some aspects of nature that we have at a given point in time. The more we understand about nature, the more we revise our theories, with the new ones representing progress and the old ones (or at least parts of the old ones) discarded as no longer adequate. We do not know whether this process will ever reach an end (i.e., whether science will eventually produce a final set of theories about how the universe works) or an asymptote (whether science will reach the limits of the human ability to understand the world), but it is clear that neither end nor asymptote has been reached thus far and that science continues to make progress, one partially wrong theory after another.

The Objectivity of Science and Science as Social Knowledge

Regardless of whether one accepts Kuhn's particular ideas about how science works (and I do only in part), we are still left with a major hurdle to reconcile the idea that science is a progressive enterprise (in the literal sense that it makes progress) and the obvious fact, repeated ad nauseam by science critics, that it is a human activity and therefore subject to the usual vagaries of all human doings. The question is: how can objectivity in science emerge by the actions of a lot of individual subjective agents?

One obvious response, most likely to be cited by scientists themselves, is that the training of a scientist, particularly in graduate school, is supposed to develop critical detachment and objectivity in its practitioners. In other words, scientists, the claim would be, are more objective than other individuals because of the way they learn and practice their profession. I don't know whether there are independent empirical verifications of this claim, but my personal experience is that it is rather unfounded. To expect a scientist to be more objective than average is the same as to expect a moral philosopher to be a saint: it may happen, but don't count on it.

Philosopher Helen Longino tackled this problem in her *Science as Social Knowledge*,[19] an attempt to strike a balance between the Scylla of viewing science as inhumanly objective and the Charybdis of complete relativism. Longino sees science as being based on evidential reasoning (i.e., reasoning about the available evidence), which in turn she treats as a human practice and therefore inherently context-dependent. This context is provided by whatever society a scientist operates within, the particular historical time, and of course the personal psychology and life history of the scientist in question. This social, as opposed to individualistic, view of science allows us to take into account social values and preferences insofar as they affect scientific research. These values and preferences, however, can be exposed and analyzed, modified or dropped, depending on whether they further or hinder the search for understanding nature that is at the core of science's mission.

For instance, we know from the study of the history of science that the rise of eugenics during the latter part of the nineteenth and the early part of the twentieth centuries was not grounded on solid evidential reasoning, but largely propelled by the racist convictions of its backers. If relativists and postmodernists were correct, there would have been no way to reasonably *argue* that eugenics is bad science, since one would simply substitute one ideology (racial equality) for another one (racism), with no way to objectively discriminate between the two. But Longino's insistence that values (and therefore ideologies) enter into science as background conditions opens up the possibility for scientists and for society at large to question the validity and effects of those conditions. Accordingly, when science blunders as it did while flirting with eugenics, both scientists and science critics play a legitimate role in correcting science—the former group by focusing on the reasonableness of the evidence available and the latter group by exposing whatever questionable values or ideologies may be playing a role in the acceptance of certain scientific claims. What is important to understand about this view of science is that there is a legitimate role for science critics, but that this role deals largely with the nontechnical aspects of science, including broad questions about what science does and why—questions to which scientists themselves are notoriously oblivious.

Another interesting implication of seeing scientific knowledge as social knowledge is that there cannot be any such thing as a lone scientist. Individuals trained in science (which is not the same as a scientist, in this context), of course, can pursue whatever question they are interested in by whatever means they have at their disposal. However, unless they communicate their findings to a larger group of other scientists (and science critics) and expose their work to both technical and ideological peer review, they are not actually doing science, according to Longino. This social dimension counters the well-known tendency of human beings to concoct stories in order to convince themselves that a particular idea is correct. It is only when we dare offer such stories to others for criticism that we begin the painful intellectual exercise that leads us to progress.

For example, consider the idea of "cold fusion." This is an alleged chemical phenomenon that would allow us to produce large amounts of cheap energy by fusing atoms at room temperature without having to use the highly complex (and energy demanding, not to mention dangerous) steps of either splitting atoms or fusing them at very high temperatures (the respective bases of the atomic and hydrogen bombs). If it sounds too good to be true, that's because it is, in fact, not true. Cold fusion was briefly in the limelight of science because of a paper published by Martin Fleischmann, a chemist, together with his former student Stanley Pons. Since this was an extraordinary claim, it immediately attracted attention, and several laboratories tried to repeat Fleischmann and Pons's results—without success. It quickly became clear that cold fusion was an illusion, and the entire field has promptly disappeared from science. It has, however, remained alive in the form of a subculture of devotees, people who remain stubbornly convinced of its possibilities and who have decided to take themselves out of the intellectual arena of established physics. The cult of cold fusion is an example of what happens when even respectable scientists with legitimate Ph.D.s are allowed to dwell only on confirmatory evidence, gingerly dismissing or rationalizing away whatever doesn't fit with their a priori conclusions. The same phenomenon is what explains why it is possible to find occasional scientists who deny global warming or reject evolution. The smaller the group of people involved, and the more they share assumptions and ideologies, the more likely science is to enter into a dead end, and the sad thing is that the people involved seem to have no sense of the amount of time and resources they are wasting in fatuous pursuit of chimaeras.

There is an additional aspect of Longino's reconstruction of science as social knowledge that is crucial, although it will be obvious to any practicing scientist: regardless of whatever social influences and assumptions enter into science, a scientist's (or anyone else's) ability to rationalize ad absurdum is limited by the fact that there is, after all, a physical world out there whose existence and behavior do not seem to depend much on our hopes and wishes. Fleischmann and Pons may have a cult following of irreducible supporters, but they will never be able to actually produce energy from an atom using a countertop apparatus simply because the laws of nature do not allow for it. As Longino puts it: "It is neither our desires, nor beliefs, nor values, nor social conventions that make a gauge register a 10 when employed in measurement but processes independent of those cultural phenomena. That we describe what has happened as attaining degree 10 rather than being intense, or having a certain density or color, or any of the other things we might say *is* [original emphasis] a social matter: a function of the language and instruments available and the kind of information we deem it important to have."[20] Longino continues:

> This view of the experience and the constraints it places on justifiable belief leads to a minimalist form of realism. There is a world independent of our senses with which those senses interact to produce our sensations and the regularities of our experience. There is "something out there" that imposes limits on what we can say about it. Once we have decided on a system for measuring

> movement, the speed of an object is not arbitrary. The sort of things we measure, however, will change as our needs, interests, and understanding change. The processes that occur in the world will continue regardless of changes in our descriptive systems.

Now that's something both scientists and science critics can reasonably rally behind.

Even though Longino's treatment of science helps mediate between relativism (science is just another cultural practice) and objectivism (science is entirely rational and its results are absolute), we need, I think, one more piece of the puzzle in order to come to a satisfactorily balanced view of how science works: a position that Ronald Giere calls "perspectivism."[21] Giere's point is that it is *both* true that science objectively expands our knowledge of the world *and* that science is a subjective activity because scientists cannot escape their limited access to the world as human beings—that is, they do not have access to a detached, all-encompassing, God's-eye view of things. Given that "objective" and "subjective" are terms usually employed in contradiction with each other, how is this possible?

Enter the intermediate solution of perspectivism. Giere's analogy is to human color vision, a phenomenon that is now understood well enough from the points of view of physics, physiology, and psychology and that can serve as an intriguing model for what it means to say that subjectivity is not necessarily in opposition with objectivity. Let us start with a classic question in philosophy, often referred to as the problem of "qualia." Qualia capture the subjective experience of phenomena such as color, the idea being that—short of getting into someone else's mind—we can never know what other people are actually experiencing when they say that they see, for instance, "red." Despite this irreducible subjectivity, however, it looks like most people can use the word "red" consistently and in a way that does seem to refer to the same sort of characteristic of objects that other people seem to refer to when they say that something is red. Indeed, when people do not use red appropriately we often can trace this inability to specific kinds of defects in their visual pathways, such as color blindness. So qualia in general, and color perception in particular, present us with a clear case where an undeniably subjective experience does not exclude what seems to be an objective assessment of the world. This problem is analogous to the one we are interested in here: the nature of science.

One way to understand the analogy is to realize that "color" simultaneously is and is not really "out there," and, similarly, both is and is not "in our heads." This sounds like sophistry, but bear with me for a moment. "Color" is a phenomenon that depends on the interaction of our sensory pathways with certain aspects of the outside world. Those aspects of the outside world are objective in the sense that they exist regardless of human intervention and can be measured by automatic instruments. In particular, the external component of color depends on the wavelengths of light as well as the reflective properties of the materials of which objects are made—none of which requires the presence of a human being. However, wavelengths and structural material properties by themselves don't generate colors. For the latter to come into the picture, so to speak, one does need a human being (or any of a variety of other animals). This is because to generate the *sensation* of color, those external components do not amount to a set of sufficient conditions: one also has to have an eyelike structure, dedicated cells that transduce the light signal, and a complex visual apparatus as part of one's brain. Any change in these latter elements will cause a change, sometimes dramatic, in the subjective perception of color, despite the fact that all the external, "objective" elements stay the same.

Color perception, therefore, is best understood as providing a particular *perspective* on the external world, one that blends objective and subjective components in ways that we can parse and understand. According to Giere, science works much in the same way, with both scientific instruments and theories playing the part of the human visual apparatus in our perception and understanding of the outside world. So, for instance, a lot of scientific instrumentation is complex, and its output depends to a large extent not just on whatever facet of the outside world the scientist is interested in measuring, but also on a series of assumptions and procedures built into the apparatus itself. When Galileo's critics were skeptical of his use of the telescope to demonstrate the existence of craters on the moon, they did have a point. After all, the Italian scientist was making extraordinary statements that, if true, would overturn centuries-old views about the cosmos, and he was doing this on the basis of an instrument that he had perfected, but that very few other people had ever used or had any confidence in. It was incumbent on Galileo to make a convincing case that the telescope was based on solid principles of optics and that it was delivering reliable magnified images not only of relatively close objects during the daytime (that was obvious) but also, more controversially, of very distant objects at nighttime. In other words, there was initially reasonable room to doubt that the new instrument was doing what Galileo claimed it was doing. After all, the Italian physicist may simply have conjured a trick or may have been fooled by some optical illusion generated by an instrument he did not fully understand. Nonetheless, Galileo and his supporters met the challenge and eventually convinced everyone else that the telescope works and that the moon's craters are real objective features of the world.

The case is similar with scientific theories, which Giere treats as perspectivist conceptual tools to understand the world. Consider evolutionary theory: it is subjective in the sense that it is a human mental construct, just like colors are; but it is also attempting to represent objective features of the world in a way that is comprehensible to human beings. It is because of the existence of such human-independent features of biology (like, for example, the fossil record) that the theory has objectively true components. The theory itself is only true in the same sense in which a normal (i.e., noncolorblind) human being "truly" sees red when looking at certain objects, since the theory is not an aspect of the world that would exist independently of human presence (fossils, on the other hand, are). This elegantly dispatches of some cherished but misleading views that scientists themselves hold about science and nature. We often hear that, as Galileo put it, "the book of nature is written in mathematical language." As I remarked before, of course there *is* no book of nature, there is only nature itself. Any theory that we construct about nature, using any kind of language, is inherently a human artifact, and it inevitably expresses a human perspective on the world. That doesn't mean it isn't true, however.

Science as a Bayesian Algorithm

I have covered a lot of complex and potentially confusing territory in the last two chapters, but I wish to conclude with a brief mention of yet another way to think about science, its power, and its limits. This is a view that is gaining favor among scientists and philosophers alike, and that I think beautifully clears up many common misunderstandings that both scientists and the general public seem to have about the nature of science, the same misunderstandings that lead people to either uncritical acceptance of science or to irrational rejection of it. I am referring to "Bayesianism," a relatively new philosophical perspective on science that has its roots in an old and, until recently, rather obscure way to think about probability that was invented by Reverend Thomas Bayes back in 1763.

Bayes realized that when we consider the probability that a hypothesis is true, we base our judgment on our previous knowledge relevant to the hypothesis in question and the phenomena it is meant to explain (i.e., we use what philosophers call inductive reasoning). We then assess new information in light of this prior knowledge and modify our belief that the hypothesis is true (or not) on the basis of both said prior knowledge and of new information. This process can be repeated indefinitely, so that the degree of trust we have in any hypothesis is always a function of the current (and ever-changing) balance between what we knew before and the new knowledge that additional data contribute.

It is easy to express Bayes's ideas in a simple mathematical formula. This is one equation that is easy to understand and that might change how you think about how science works and, more generally, how you assess your beliefs. The equation is known as Bayes's rule, and it looks like this:

$$P(H|D) = \frac{P(D|H) \times P(H)}{[P(D|H) \times P(H)] + [P(D|{\sim}H) \times P({\sim}H)]}$$

Where $P(H|D)$ (which reads, "the probability of H, given D") is the probability that the hypothesis is correct given the available data; $P(D|H)$ is the probability that the data will be observed if the hypothesis is correct; $P(H)$ is the unconditional probability of the hypothesis (i.e., its probability before we knew of the new information). The denominator repeats the numerator with the addition of an equivalent term that includes the probability to observe the data if the hypothesis is wrong and the unconditional probability that the hypothesis is wrong. (In the more frequent case of multiple hypotheses being considered at the same time, the denominator includes the probabilities of observing the data given each additional hypothesis, as well as the unconditional probabilities attached to each alternative hypothesis; the ~ symbol stands for *not.*) The denominator of the right-hand side of the equation is also known as the *likelihood* of all hypotheses being considered. The left-hand side of the equation is called the *posterior probability* of the hypothesis being examined; the left-hand part of the numerator on the right side of the equation, $P(D|H)$, is known as the *conditional likelihood* of the hypothesis in question; and the right-hand part of the same numerator, $P(H)$, is called the *prior probability* of the hypothesis being considered. This sounds very complicated until we examine a particular example, so bear with me for a few more paragraphs.[22]

A family has plans to go fishing on a Sunday afternoon, but their plans depend on the weather at noon on Sunday: if it is sunny, there is a 90 percent chance they will go fishing; if it is cloudy, the probability that they will go fishing drops to 50 percent; and if it is raining, the chance of going drops to 15 percent (these probability estimates need not be precise for the system to work). The weather prediction at the point when we first consider the situation calls for a 10 percent chance of rain, a 25 percent chance of clouds, and a 65 percent chance of sunshine. The question is: given that we know the family eventually did go fishing (the data), was the weather sunny, cloudy, or rainy (the three hypotheses)? You will probably have your intuitions about this, and they may well be correct. But science goes beyond intuition to empirically based reasoning. Here is how Bayes would solve the problem:

Let's plug our preliminary assessment of the situation into Bayes's rule:

- The probability of fishing (F) if it is sunny (S), $P(F \mid S) = 0.90$
- The probability of fishing if it is cloudy (C), $P(F \mid C) = 0.50$
- The probability of fishing if it is rainy (R), $P(F \mid R) = 0.15$

The probability of each kind of weather, given the predictions of the weather report, can be summarized as follows:

- The probability of sunny weather, $P(S) = 0.65$
- The probability of cloudy weather, $P(C) = 0.25$
- The probability of rainy weather, $P(R) = 0.10$

Notice that the sum of the probabilities of each weather condition is 1 (i.e., 100 percent): $P(S) + P(C) + P(R) = 0.65 + 0.25 + 0.10 = 1.00$. Also note that these hypotheses are mutually exclusive in the sense that it was either sunny or cloudy or rainy, but not a combination of these. The overall likelihood of going fishing (the denominator of the right side of Bayes's rule), $P(F)$ is 0.725: $[P(F \mid S) \times P(S)] + [P(F \mid C) \times P(C)] + [P(F \mid R) \times P(R)] = (0.90 \times 0.65) + (0.50 \times 0.25) + (0.15 \times 0.10)$.

We can now obtain the updated conclusions about our hypotheses on the weather, given the prior and new information (the latter being that the family *did* go fishing). So given that the family went fishing, according to Bayes's rule,

The probability that the weather was sunny is as follows:

$$P(S \mid F) = [P(F \mid S) \times P(S)] / P(F) = (0.90 \times 0.65) / 0.725 = 0.807$$

The probability that the weather was cloudy is as follows:

$$P(C \mid F) = [P(F \mid C) \times P(C)] / P(F) = (0.50 \times 0.25) / 0.725 = 0.172$$

And the probability that the weather was rainy is as follows:

$$P(R \mid F) = [P(F \mid R) \times P(R)] / P(F) = (0.15 \times 0.10) / 0.725 = 0.021$$

Finally, note that $P(S \mid F) + P(C \mid F) + P(R \mid F) = 0.807 + 0.172 + 0.021 = 1.00$, because one of the hypotheses *must* be true (it was sunny or cloudy or rainy; no other possibilities are in the mix).

Bayes's rule, therefore, tells us that—given the prior knowledge of the situation we had and the new information that the family went fishing—the likelihood that the weather was sunny was the highest. Well, you could have guessed that, right? Yes, in this simple case. But notice that Bayes's rule gives you additional information: First, it tells you what the *best* available estimate of the probabilities of all three hypotheses is. Consequently, it tells you how confident you can be that the weather was sunny (which is better than simply saying "it's more likely"). Also, it is clear from the equations that the probability of the hypothesis that the weather was sunny *increased* with the new information (from 0.65 to 0.807), which should increase your confidence that the sunny hypothesis is correct. Finally, Bayes's theorem reminds us that our degree of confidence in any hypothesis is never either 0 or 100 percent, although it can come very close to those extremes—scientific hypotheses are always tentative.

Bayesian analysis is a good metaphor (some philosophers of science would say a good *description*) of how science really works, a position best referred to as Bayesian*ism* (to distinguish the philosophy from the theory of probability). Moreover, it is a good description of how *any* rational inquiry into the world goes, if it is based on a combination of hypotheses and data. The scientist (and in general the rationally thinking person) is always evaluating several hypotheses on the basis of her previous understanding and knowledge on the one hand, and of new information gathered by observation or experiment on the other hand. Her judgment of the validity of a theory therefore changes constantly, although very rarely does it do so in a dramatic fashion. Bayesian analysis thus clearly shows why it is irrational to hold to extreme positions in most empirical matters: if your priors are 0 or 1, your conclusions are impervious to new information, and they will not change *no matter what the data say*; that is, you take them on faith.

When we combine sensible ideas about social criticism of science with those about paradigm shifts, perspectivism, and Bayesian operations, we arrive at a reasonable view of science: yes, it is a subjective human activity affected by psychological and societal vagaries; but, no, it is not an arbitrary activity, and it does make progress, albeit always tentatively and in a way that is constantly subject to the possibility of revision. This, at least, is what we can glean from what experts in science, sociology, and philosophy tell us. One more question, however, remains to be addressed before we can arrive at a satisfactory conclusion of our quest about the nature of science: why should we trust experts to begin with? To this issue we turn in the next chapter.

CHAPTER TWELVE

Who's Your Expert?

For every expert there is an equal and opposite expert; but for every fact there is not necessarily an equal and opposite fact.—Thomas Sowell, American economist

I have as much authority as the Pope, I just don't have as many people who believe it.—George Carlin, American comedian

Why should you trust anything you've read so far in this book? Certainly not because it is printed black on white, as there are plenty of books out there that are not worth the paper they are printed on. Nor because I peppered the text with scholarly looking endnotes, since Bjørn Lomborg's *The Skeptical Environmentalist*, as we have seen, is full of notes (2,390, to be precise) and yet amounts to a highly misleading treatment of the problem of global warming. It isn't enough that I have a Ph.D. after my name, since Michael Behe of Lehigh University also has one, and yet he would have you believe that astrology and intelligent design are legitimate sciences. Conversely, plenty of reasonable, informative, and insightful books are written by people with no academic degrees at all. However, you will likely not have the time to fact-check every assertion I made in this book or to read it carefully enough to analyze the fine structure of my arguments, searching for logical fallacies or weak links.

This problem is arguably a major issue underlying the debates about science in the public arena, and it boils down to questions of expertise and trust. What qualifies someone as an expert? And what degree of trust should we put in what that person says about a given topic? We saw earlier that a large portion of the media seems to have adopted the idea that whenever controversial issues are discussed, it should not take sides, but instead present a "balanced" treatment, ideally giving equal time to "both sides." But often there are more than two sides, and not infrequently one side is downright silly or plainly wrong, despite the fact that it is not too difficult to marshal "experts" to defend its positions. We also found that an alleged expert, the above-mentioned Lomborg, got it dramatically wrong on global warming, while a politician with no technical expertise in the field, Gore, got it (largely) right. When we analyzed the Dover case on intelligent design, we discovered that Judge Jones, with no expertise in either science or philosophy, delivered a stunningly sophisticated opinion, displaying an intricate and pertinent understanding of the philosophy of science that was presented at the trial through the briefings submitted by the various experts. Finally, we discovered that while intellectuals are under assault in modern society, there has been a worrisome commodification of "expertise" through the phenomenon of the so-called think tanks. What is the average intelligent citizen to do?

The problem of expertise is not a new one. Plato tackles it in one of his Socratic dialogues, the *Charmides*, in which Socrates is engaged in a discussion of what constitutes temperance. At one point, the philosopher takes a comment by one of his interlocutors as a reason to explore the relation between science and medicine, which in turns leads him to inquire as to how someone might be able to tell good from bad medicine. The *Charmides* is a strange early Platonic dialogue, and scholars are not even sure what the subject matter of the dialogue actually is.[1] Nonetheless, there is a small section of the dialogue that is pertinent to our inquiry, as Socrates wishes to explore the question of how "a wise man" can possibly distinguish a doctor from a quack (notice that, as usual, this is not really a dialogue so much as Socrates—who does most of the talking—pursuing a point with the assistance of a willing interlocutor who simply nods along the way):

SOCRATES: "Let us consider the matter in this way. If the wise man or any other man wants to distinguish the true physician from the false, how will he proceed? . . . He who would inquire into the nature of medicine must test it in health and disease, which are the sphere of medicine, and not in what is extraneous and is not its sphere?

OTHER GUY: *True.*

SOCRATES: And he who wishes to make a fair test of the physician as a physician will test him in what relates to these?

OTHER GUY: *He will.*

SOCRATES: He will consider whether what he says is true, and whether what he does is right, in relation to health and disease?

OTHER GUY: *He will.*

SOCRATES: But can anyone pursue the inquiry into either unless he has a knowledge of medicine?

OTHER GUY: *He cannot.*

SOCRATES: No one at all, it would seem, except the physician can have this knowledge—and therefore not the wise man. He would have to be a physician as well as a wise man.

OTHER GUY: *Very true.*[2]

That, in a nutshell, is the problem of expertise (to which Socrates, by the way, did not have a solution): it would seem that the only way to assess whether an expert is, in fact, an expert is to be an expert oneself. On the positive side, this is what justifies peer review in science: if I wish to write a paper in quantum mechanics, say, or on evolutionary theory, I need to submit the paper to the (usually anonymous) review of several scientists whose specialty is quantum mechanics, or evolutionary theory. The idea, of course, is not that experts are infallible, but rather that they are the best (often the only) people qualified to make a judgment on technical matters.

On the negative side, however, it would seem that if I think (and I do) that Deepak Chopra talks nonsense when he tells people about quantum mechanical elixirs of youth, I would first have to be an expert in quantum mysticism. But the problem is that quantum mysticism is (I think) quackery, and that therefore there is *no such thing* as an "expert" on quantum mysticism. This despite the fact that quantum mechanics is one of the best-established scientific theories of all time (then again, I'm not an expert in quantum mechanics . . .). This same conundrum underpins much of what I have written in this book: how is the average intelligent person (Socrates' "wise man") supposed to be able to distinguish between science and pseudoscience without becoming an expert in both? Moreover, how is it even possible to become an expert in nonsense?

Research on Expertise: Asking the Experts

As it turns out, much scientific research has been done on expertise, so in some sense we can say that there are experts on expertise. It may be instructive to take a look at what they found. For instance, Anders Ericcson and Jacqui Smith begin their discussion of expertise with two conclusions from their work: first, a single success, as spectacular as it might be, does not make one an expert. Second, being an expert in a given field has little to do with one's general abilities to retain information.[3] The first point is not exactly surprising: someone can pull off a once-in-a-lifetime success, like penning a bestselling novel or writing a song that makes it to the top of the charts, without necessarily being an expert in how to write bestselling books or songs. On the other hand, few people win the Nobel more than once, yet it is hard to conceive of an amateur achieving such an honor in today's sophisticated and specialized world, regardless of whether the discipline in which the Nobel is bestowed is a scientific one (where the expectation of expertise is almost a given) or not. The point about memory, however, is a bit more subtle: while a good general memory certainly will not hamper anyone, studies have shown that experts in a field do not have better general memory than novices; rather, the difference is in the way the expert *structures* her memories of the *pertinent* knowledge. In this sense, experts seem to work like the appropriately named "expert systems" in computer science. Expert systems are not just databases of information that can be retrieved by brute force. Rather, the software focuses only on a particular domain (of expertise) and gains its power from the fact that the programmers organize that special knowledge in ways that reflect our best understanding of whatever the specific field happens to be, medicine, law, engineering, or something else.

According to research summarized by Ericcson and Smith, not surprisingly, it takes time to become an expert. Studies of chess have become particularly useful in understanding expertise, since the game is characterized by a small number of rules and yet also by an open-ended universe of possible outcomes, making it a good model system for social scientists. Most studies on chess distinguish novices from expert players, and the latter from masters. The results show that it takes about three thousand hours of playing experience to become an expert, but about ten times as much to qualify as a master. Interestingly, the rule of thumb that a decade or so is needed to become an expert in a field applies not only to chess playing, but also to science and arts. A typical doctoral degree in the sciences takes four to six years in the United States, to which one has to add four years of college, and increasingly an additional number of years as a postdoctoral associate. It is not unusual to begin a tenure-track academic job after one has already spent a decade learning about specialized aspects of the field. But expertise is about seeing patterns, not sheer amount of knowledge. In the particular case of chess, according to Ericcson and Smith, and contrary to popular belief, masters don't actually calculate a much larger number of moves ahead when compared to novices; instead, their experience allows them to identify patterns on the chessboard that suggest which move(s) will be most profitable, given the current state of the system.

Based on research conducted over the span of decades, Ericcson and Smith propose a three-stage process for the acquisition of expertise: at the initial, cognitive, stage, a person focuses on understanding what the task actually is, trying to discriminate between relevant and irrelevant information. This is the situation of an amateur chess player or a beginning undergraduate student. The second stage, the associative one, is characterized by quantitative improvements in the ability to retrieve information, largely because such information gets filed in a structured way, with associations being noticed and built among disparate pieces of information. A chess expert (but not yet a master) or a graduate student may be at this stage. Finally, we have the autonomous stage, where performance becomes largely automatic, with little actual conscious thought going into most of the tasks at hand. A chess master or a professional academic are at this level. Speaking from personal experience, one develops a "feeling" for what happens, an intuitive sense of what is relevant information and what constitutes background noise, or for what will likely turn out to be a good approach to a problem, as opposed to a waste of time. Of course, the expert can reflect on her intuitions and articulate *reasons* for what she is doing, but the answer comes "in a flash," as they say, not as the result of hours of explicit logical pondering.[4]

As we saw above, questioning what constitutes expertise began (as far as we know) with Plato's inquiry into the difference between a doctor and a quack. It is appropriate, therefore, to look at what Barbara Daley has found out about expertise in the medical profession, with particular respect to nurses.[5] The results paint a picture that is remarkably parallel to the one emerging from the study of chess playing, despite the obvious dissimilarities between the two fields. Again, generalized memory has little to do with it, because even skilled professionals' ability to recall raw information is pretty limited; what the experts do instead is they use their accumulated structural knowledge of what works and doesn't work as they have encountered it in their experience. Expert nurses, unlike novices, see things more systemically, being aware of a variety of issues including organizational structure, availability of resources, and even politics. The experts augment their learning through informal encounters, and they rely much more on their own sense of what they need to learn. Beginners, on the other hand, need more structured, formal education, and, in the words of one of the interviewed nurses, they "did not even know what they did not know,"[6] or "what I didn't know as a novice, and what I learned in my career, was how to learn."[7]

Similar results were obtained by Gaea Leinhardt in a study of teaching techniques as implemented by novice and expert math teachers.[8] Notice that these are people who all understand math very well; what they differ in is their experience in *teaching* math. As I mentioned earlier in the book, Murrey Sperber denounced in his *Beer and Circus: How Big Time College Sports Has Crippled Undergraduate Education* what he called the "good researcher = good teacher" myth propagated by both faculty and administrators at universities worldwide. Sperber contends that there is not a shred of empirical evidence for the somewhat intuitive idea that because someone understands a subject very well, even excellently well, one is therefore capable of explaining (teaching) it well. Leinhardt's results lend empirical foundation to Sperber's contention (as well as to my personal experience in academia). Novices who had a similar background in math to that of experts but had much less teaching experience structured their lesson plans in a superficial and ineffective way, demonstrating lack of depth in their ability to connect concepts, to come up with innovative examples, and to build on previously taught ideas. As Leinhardt put it, "teachers seem to grow [with experience] from being structured and somewhat rigid to being introspective, flexible, and responsive."[9] Expertise means familiarity with a complex territory that is a confusing terra incognita for nonexperts.

Another author who has written on the nature of expertise, drawing from a variety of empirical studies, is James Shanteau.[10] His conclusion is not only that novices and experts use about the same number of "clues" to arrive at their decisions (and a small number at that: often between one or two and six or seven pieces of information), but also that novices have a tendency to use *too much* information, getting stuck because of their inability to discriminate between what constitutes a useful clue and what doesn't. The clues used by experts to arrive at a particular decision are more informative than those used by novices, because experts have a conceptual model of the field that makes use of natural correlations among pieces of information: when information is structured, a small number of items may actually embody a significant wealth of detail that can be read by someone who knows what to do with it. Unfortunately, concludes Shanteau, it may take a lifetime to figure out which variables to pay attention to and which to ignore. As Philip Ross points out, moreover, expertise does not translate from one field to another, not even across closely related fields.[11] Ross cites research conducted by psychologist Edward Thorndike, who a century ago discovered that studying Latin did not improve one's understanding of another language, such as English; nor did proficiency in geometry improve one's performance in logic.[12] This also explains what I call the "Nobel effect." It is unfortunately not uncommon for people who win the Nobel (in whatever area, but especially in the sciences) to begin to pontificate about all sorts of subjects about which they know next to nothing. Winning a Nobel because of research in, say, quantum mechanics requires a large amount of expertise in quantum mechanics. But that expertise simply does not carry over to issues of international politics or economics, or even to other fields of science. This is why creationist claims that "legitimate scientists" oppose evolution are empty: a physicist or chemist is no more qualified to comment on a biological theory than a biologist is qualified to comment on quantum mechanics or the theory of chemical bonds.

Time, of course, is a necessary but not sufficient condition to become an expert: Ross refers us to studies by Anders Ericcson showing that to become an expert, one needs to engage in "effortful study." We all know that an amateur musician, say, may spend thousands of hours practicing an instrument without ever getting past a relatively early stage of development in his musicianship. In order to get to the next level, one needs to challenge himself with difficult things that one cannot, at the moment, do. This once again fits with the idea that expertise is not a matter of accumulating raw information or simple experience, but requires engaging with the tasks at hand in order to develop and internalize heuristics that help us solve problems. This is also why the heuristics we develop in one field usually do not apply to another one (although the ability to *learn* may, in fact, be what psychologists call a "portable skill").

Nonsense on Expertise: The Postmodern Critique

It would be surprising if a topic so potentially controversial as that of expertise were to be left alone by our old friends, the postmodernists. Sure enough, the same postmodernists who have declared the death of the author (despite the rather surprising simultaneous survival of the text) have also declared the (metaphorical) death of experts, using pretty much the same "logic" we examined in the last chapter. As an example, let's take a look at an article by Leroy White and Ann Taket.[13] While the paper focuses on expertise (or lack thereof) in operational research (a field of study on how to use analytical methods to make better decisions), the arguments proposed by White and Taket are quite general and will therefore serve as a further case study in how the postmodern mind (doesn't) work. As the reader will see, I will have to quote extensively from White and Taket's article. This is not because their prose is particularly worthy of direct quotation or because I would otherwise be unable to paraphrase their meaning, such as it is. It is just that anything other than direct and precise quotations would likely elicit a sense of disbelief in the reader, leading him to conclude that I'm simply making the stuff up. I am not. It's too good to be able to surpass it with mere fiction.

White and Taket begin their paper by calling for a different approach to expertise, one that "does not privilege authority, rationality or scientific discourse." Indeed, they conclude their piece several pages later with, "We are overwhelmed with the stench of rationality, reason and privilege." Fighting words, for sure. Now most people would agree that "authority" per se is not a good guide to anything. After all, as anyone who has taken Philosophy 101 should know, an argument from authority is a classic logical fallacy. In fact, this is precisely the heart of the problem of expertise: how do we tell reliable authorities (experts) from phonies? How do we figure out who is the real McCoy? But rejecting rationality and science seems, well, irrational, on the face of it. It is one thing to caution against relying too much on science, especially for solutions to problems that may not, by their nature, be particularly amenable to scientific treatment. But an outright rejection of anything to do with science and rationality, to the point of invoking their (presumably metaphorical) "stench" is one telling sign that we are not dealing with a serious intellectual point. And yet, there is more, much more.

Again, White and Taket: "As postmodernists, our view of the world is text-centred. Everything can be seen as text. . . . All phenomena and events can be regarded as text. Hence, a meeting, a war, a holiday, buying a car, speech, and so on, can all be assigned the order of text. Furthermore, these texts are also inter-connected, in that each system of signs can be transferred into others. . . . An endless connection between texts with no prospect of ever arriving at an agreed point; thus, meaning is endlessly deferred. Everything is related to everything else."[14] I have news for White and Taket: everything is *not* related to everything else. That extreme position is the stuff of comic novels, such as Douglas Adams's *Dirk Gently's Holistic Detective Agency* series, but not of serious scholarship.[15] If "everything" is text, then the word "text" loses meaning. A war and a holiday may have something in common (they are both human activities, for instance), but that something in common is simply too general to get much mileage out of it. Following the postmodernist route we may indeed never arrive at meaning, but not because meaning is not there, only because we are lost in endless linguistic games that are entirely beside the point.

White and Taket claim to be performing nothing short of an "autopsy of the modern expert," and one may wonder what qualifies them to do so. Not to worry, we find their position about their own work on the second page of the paper: "It might appear that this paper is tending to write authoritatively about our aims. We would like to say that this is a postmodern text and that no such reading should be sought; as authors we do not wish to be authorities in its interpretation." I warned you I wasn't making this stuff up! What does it *mean* for the authors of a text (an article really does qualify as text, and those who wrote it really are authors) to tell their readers that the authors themselves do not have any authority in the interpretation of what they wrote? Did they write random gibberish with no intention to communicate a specific message? It appears not, since White and Taket are very specific about what they want to say and why. But they better take *some* responsibility for it, or one begins to question what the point of the whole exercise actually is.

Indeed, it is responsibility itself with which White and Taket have a problem. "[E]xperts can neither claim causal agency nor imply social responsibility. The expert, as causal agent, who is the author(ity) on cause, privileges him(her)self or sets him(her)self aside from the cause itself. Such strong claims on causality become untenable if we note that everything is related to everything else . . . In this postmodern world, experts are not allowed to take the credit nor hold responsibility for disasters. They are a tiny part of a complex exchange . . . This means an end to the idea of an individual moral agency."[16] OK, let me get this straight: since, according to White and Taket, everything is connected to everything else (which is not true), then any given individual cannot play a significant causal role in anything (presumably, because causality is distributed throughout the universe), and therefore nobody can take credit or blame for anything. The implications of this sort of reasoning for society would be staggering if anyone but a few misguided souls actually took it seriously. Indeed, I'm pretty sure one can make the case that not even White and Taket truly believe what they are saying. For instance, suppose the authors live in an apartment building that was built by a particular construction company, following the plans laid out by a particular group of engineers. Now let us suppose further that the building suddenly collapses, and that the subsequent investigation of what happens pins responsibility squarely on the shoulders of the engineers, who in a flagrant effort to save a few dollars had produced a technically flawed design. Are we to understand that White and Taket would forgo their right to sue because "everything is connected to everything else" and therefore nobody can be blamed for anything? I doubt it.

Then again, maybe they wouldn't sue. In their paper they actually examine a case involving engineering:

> We can say that in the construction of a bridge, a certain scientific knowledge is deployed and it may require a certain "applied science" to provide data on the details of the bridge's construction. But, the creation of a bridge is a design exercise which is not wholly based on scientific data. It is based on its functions, with associated costs, which would have to be balanced with uncertainty and conflicting values. It is not just a question of optimizing a model, further complications will arise; for example, concerns of local residents and issues such as displacement, the environment and so on.[17]

Notice the use of scare quotes around "applied science," as if the authors only grudgingly acknowledge the existence of any form of scientific expertise. As with much postmodernist literature, what they write in this instance is either trivially true or outrageously false. The trivially true interpretation is that of course there is more than engineering to the construction of a bridge: there are, in fact, plenty of social, economic, and cultural issues coalescing to form the broad picture. But whether the bridge is well designed or not is *solely* a matter of engineering, and the rest simply doesn't enter into it. This implies that if the bridge collapses because of a design flaw, the engineers are both legally and morally responsible for the disaster. The alternative interpretation of what White and Taket are saying, however, denies this very conclusion, claiming that somehow the fact that bridges are built not just through engineering, but also as the outcome of a variety of societal factors, absolves engineers of any responsibility (and, for that matter, denies them any praise) for their work. This is nonsense on stilts in the utmost degree.

White and Taket, like so many postmodernists, clearly have a problem with science, presumably because science is often considered an "authority," and—to be sure—not always for the right reasons. But it is one thing to level a measured criticism aimed at the abuses of science and quite another to call for an outright rejection of the whole practice because one is in principle against authority tout court. "The expert has no authority over the text [remember that "text" here means pretty much anything and everything produced by human beings]. No postmodernist would really believe that the expert has the right answers, only that they act (consciously or unconsciously) to monopolize authority . . . Insisting on the impartiality of expertise may be a pretence for imposing 'a view,' thereby making the expert an instrument of power."[18] Notice here the use of a common logical fallacy, that of introducing a false dichotomy. According to White and Taket, there are only two possibilities: either experts have the right answers or they are simply acting to impose their view in a quest for power. For a postmodernist there cannot be any such thing as a "right" answer, so, given the two choices, it must be the latter. But the real world is more complicated than that. Experts often *do* have a good answer to a given question, but not always. Sometimes there may be more than one possible answer. Moreover, the expert could be *both* right *and* seeking to impose a particular opinion out of conscious or unconscious motives. But to dismiss someone's arguments on the basis of his motives constitutes yet another logical fallacy, known as the genetic fallacy. Motives, as important and perhaps at times suggestive as they may be, still ought to be considered separately from the *reasons* supplied for a given answer.

This, shall we say, irrational disdain for rationality reaches its apotheosis in White and Taket's paper when they write that "following Lyotard, we would contest this view [of the existence of a logic characteristic of operational research techniques] believing that they have no special logic, and that they have the same status as myths, witchcraft, religion and science. We also view rational explanations as a product of the Enlightenment, and as such are guilty of all the errors attributed to it, of which there is a long history. Rational approaches are dominating and oppressive, and will always be brutal to someone or other."[19] First off, notice that myths, witchcraft, religion, *and science* are put in the same dustbin. Yet I am willing to bet that, when sick, most postmodernists will rely on medical science, not on witchcraft. Second, rational explanations, although central to the project of the Enlightenment, were most emphatically *not* an invention of that cultural movement: Western philosophy from the pre-Socratics, and especially Plato, onward has always been about rational explanations. And why, one might ask, should rationality be oppressive and brutal, as opposed to liberating and helpful? Because, as White and Taket inform us a few lines later, "words, reason and rationality are inconsistent with emotions, feelings, creativity, and so on." Inconsistent? Certainly reason is not the same thing as emotion, but philosophers ever since David Hume and modern neuroscientists like Antonio Damasio have convincingly argued that reason and emotions are both essential components of being human and that they need to be brought into balance for a human being to function properly.

Hume famously said that "we speak not strictly and philosophically when we talk of the combat of passion and of reason. Reason is, and ought only to be the slave of the passions, and can never pretend to any other office than to serve and obey them."[20] A surprising stance for a philosopher of the much maligned Enlightenment! What Hume meant was that reason may provide us with the most effective means to achieve a particular end, but we *care* to reach that end because of a "passion" that we hold and that motivates us. Similarly, modern neurobiological research has brought Damasio to conclude that the very essence of human consciousness has its roots in a continuous balance between our emotions and our ability to reason.[21] We can now map which areas of the brain are involved in rational decision making (largely the frontal lobes), and show that they are intricately interconnected with the neurological seat of the emotions (primarily the amygdala). But of course all of this assumes that you think that modern science is a few notches more sophisticated than witchcraft.

From the postmodernist's standpoint, the very concept of expertise is a recent, artificial creation: "Following the project of Foucault, others have shown that the concept of expertise is not a natural one, in that it only erupted in certain fields at particular points in history, and its form is unstable . . . the expert is a modern figure, emerging in the 19th Century, a creation of liberal discourses, who plays an integral part in governing and controlling individuals and populations at a distance, through the establishment of a network of experts."[22] Really? Let us leave aside the final paranoid statement conjuring up a worldwide cabala of experts (in what?) bent on world domination. Foucault and company are simply wrong to trace the concept of expertise or the social role of experts to the nineteenth century: as we saw at the beginning of this chapter, Socrates posed the question of what makes an expert twenty-five centuries ago, well before modern "liberal discourse" and presumably because people even then did consider—rightly or wrongly—some fellow human beings as experts in, say, medicine. Moreover, what does it mean to dismiss expertise on the grounds that "it is not natural"? Neither is the computer on which I am writing this chapter or, I'm pretty sure, whatever instrument White and Taket used to write their essay. I'm beginning to sound repetitive, but we have yet another example of a previously encountered logical fallacy: the naturalistic one, or the equation of what is natural with what is good (how about poisonous mushrooms?).

Perhaps the most paradoxical part of White and Taket's article, though, is when they actually attempt to do what one would think is contrary to their whole point: they give us a definition of a postmodern expert: "Following Funtowicz and Ravetz, the 'postmodern expert' (!) is a person 'who happens to be able to have an experience'" ["!" in the original].[23] Being able to have an experience qualifies one as an expert? No wonder later on they conclude that "a postmodern [operational research program] might appear interesting and fascinating. It might also appear ridiculous and absurd. We make no attempt here to distinguish between them."[24] Of course not. If one accepts the postmodernist worldview, on what basis could such distinction possibly be made?

Back to Reality: Once Again, How Do We Tell an Expert from a Phony?

As we have seen, one can do really unsound philosophy while talking about expertise, as in the case of White and Taket. However, the question is not wholly empirical either: while scientific studies of expertise such as those discussed early on in this chapter are certainly valuable and often intriguing, they cannot settle the question of most concern to us here: how is the person on the street (so called) supposed to tell whether an alleged expert is right? Moreover, how should we make up our minds when experts seem to disagree on the matter at hand? In this book we have faced precisely this situation several times: are we to go with biochemist Michael Behe, a supporter of intelligent design, or with cell biologist Ken Miller, a supporter of evolution? After all, both have legitimate Ph.D.s, both teach at respected universities, and both have published in peer-review journals. Should we conclude from this that the matter is so much up in the air that at this moment there is no reasonable way to take sides in the debate between creationism and evolution? If the reader has made it this far, she knows that my answer is most definitely negative: Behe is spectacularly wrong, and Miller is (largely, but not entirely) right. But how do *I* know?

Some interesting light on this question can be shed by the philosophical analysis of expertise provided by Alvin Goldman.[25] Goldman begins by departing from the classical philosophical approach to epistemological problems such as this one. He rightly says that in classical epistemology one makes the assumption that all agents are perfectly rational, and that they have "unlimited logical competence." This is somewhat similar to the simplistic assumption underlying much classic economic theory, namely, that market agents are perfectly rational and have access to all the relevant information about the companies, resources, capitals, and other factors that comprise the market. But just like markets are most certainly not made of perfectly rational agents (which is why economic predictions have a spectacular tendency to fail, as we have been painfully reminded in recent times), neither can we assume that nonexperts in a field are characterized by perfect logical competence to assess all the relevant facts. So what is an imperfect "epistemic agent" to do? Goldman's answer is that it is largely evidence about *the speaker* (or writer), instead of evidence about the matter at hand, that allows us to reasonably make up our mind on whether, say, Behe or Miller are more likely to be right. That's because, as Goldman puts it, "the novice either has no opinions in the target domain, or does not have enough confidence in his opinions in this domain."[26] Or, I would add, *shouldn't* have confidence in his opinions, precisely because he is no expert on the matter and the matter is intricate enough to require expertise.

Before proceeding with his recommendations on how to judge whether an expert is likely to be on track or not, we need to take a look at Goldman's very definition of expertise, which is intuitively appealing and provides in itself some useful insight into the issue. Goldman says that expertise is a question of "veritistic terms," philosophical jargon for truth-linked assertions. In other words, someone can be considered an expert in a particular domain if two conditions are met: (1) that person holds to more true beliefs (and fewer false beliefs) about that domain than a nonexpert; (2) that person holds a "substantial body of truths" in that domain. This may seem a bit vague, but it implies two facts that intuitively one would want to be part of the idea of expertise: the first condition means that, in order to qualify as an expert, someone has to know things *better* than most people when it comes to a particular domain. If I claim to be an expert in baseball (I am not), I ought to get a lot more things right about all aspects of baseball, and fewer things wrong, when compared to someone who admits he is no baseball expert. But that is not enough, because one could know more than someone else and still not know nearly enough to be an expert (one could be a good amateur or just have an occasional interest in the matter). Hence the second condition, which insures that there is a *substantial* body of domain-specific knowledge in the toolkit of the alleged expert.

Let us now turn to the core of Goldman's proposal, his five-point guide to figuring out whether your trust in a prospective expert is well founded or not. The five kinds of evidence that a novice can use to determine whether someone is a trustworthy expert are:

- an examination of the arguments presented by the expert and his rival(s);
- evidence of agreement by other experts;
- some independent evidence that the expert is, indeed, an expert;
- an investigation into what biases the expert may have concerning the question at hand;
- the track record of the expert.[27]

Every so often all one needs is the first criterion: we, as nonexperts, are sometimes able to tell whether an alleged expert's arguments hold up to even superficial scrutiny (what Goldman calls "direct argumentative justification"), and we may be able to tell he is a phony even if we are not capable of directly refuting his arguments (Goldman's "indirect argumentative justification"). For instance, a politician is—under ideal circumstances—an expert in public policy, someone who wishes to convince us that he has a better idea than his opponents about how to run the "public thing" (as the ancient Romans used to call the state). It isn't as easy as one might wish to spot a phony politician (or simply a politician whose ideas are confused or wrong), because most members of the public do not know enough specifics about economics, foreign affairs, law, etc. to make a judgment by direct justification. To complicate things, judgments by indirect justification, for instance based on how articulate the politician in question is, are suspect because the art of politics is closely allied to the art of rhetoric, so effectively criticized by philosophers ever since Socrates.[28]

Nonetheless, an informed public may still be able to make inroads about the reliability of a particular politician by using both components of Goldman's first criterion. To begin with, most politicians are actually *not* more knowledgeable about, say, economics or foreign affairs, than the rest of us. This means that several of their arguments can be evaluated directly, if only one bothers to pay attention and inform oneself. Moreover, politicians can sometimes be caught in flagrant contradictions on the basis of what they say to different audiences, which means that even an indirect analysis may reveal that there is something fishy going on. The problem with Goldman's first approach comes when we turn to more esoteric realms, such as the debates on evolution or global warming that we considered previously. There, most people really do not have the technical background necessary for a direct evaluation of the arguments, and indirect techniques often fail because, for instance, a creationist evangelical preacher is likely to be much more rhetorically skilled than an academic professor who rarely leaves the ivory tower. While there is much to be said for encouraging academics to get out into the real world and talk to people, they are simply not trained in rhetoric, and they usually (though not always) are unable to effectively rebut so-called commonsense arguments advanced by the other side.

We are then onto Goldman's second criterion: look for evidence of agreement among experts. It works like this: you may be convinced that your car doesn't start because of a problem with the carburetor, but if several car mechanics tell you that your carburetor is fine and that instead you need to replace, say, the injection pump, you'd be a fool if you went ahead and bought a new carburetor. However, as we have seen when talking about both evolution and global warming, it is all too easy to find at least *some* "experts" who will defend almost any sort of nonsense. The annals of science are littered with (sometimes even prominent) scientists being taken in by one crackpot idea or another. Just think of Newton's fascination with alchemy or Alfred Russel Wallace's (the codiscoverer with Darwin of natural selection) defense of spiritualism. This is why creationists and supporters of intelligent design play the game of gathering signatures from "experts" who criticize evolutionary theory. Often such criticisms are phrased in such generic terms that one would be hard pressed *not* to sign on even if one were a hardcore evolutionary biologist. For instance, scientists sometimes are asked to endorse statements to the effect that critical examination of scientific theories ought to be taught to high school students; what scientist could possibly object to the use of critical thinking in schools? But the real issue is why the authors of such petitions usually single out only the theory of evolution for scrutiny, and not the theory of gravity or the germ theory of disease, for example. Why the focus on evolution, if there really is no ideological agenda propelling them?

Goldman's third criterion may be helpful here: look at the independent evidence that the expert is, in fact, an expert. That immediately rules out most creationists and ID supporters: the overwhelming majority of them do not have degrees in the sciences, and some of them—like the infamous "Dr." Kent (Dino) Hovind, whom I've had the dubious honor of debating in public more than once—actually have "degrees" from unaccredited universities (Hovind sports a "Ph.D." in Bible studies from Patriot University, which turns out to be someone's house). Even those that do have degrees, such as Duane Gish of the Institute for Creation Research or Bill Dembski of the Discovery Institute, have acquired them in unrelated fields (biochemistry and philosophy, respectively). A Ph.D. in biochemistry does not make someone an expert in evolutionary biology any more than my Ph.D. in evolutionary biology makes me an expert in biochemistry (trust me, I'm not). This still leaves a small number of experts with good credentials (graduate degrees in the appropriate discipline from accredited universities) who turn out to be creationists. But here we can go back to Goldman's second criterion and follow his suggestion to survey the field: if the overwhelming majority of accredited experts sides one way, that way is the most likely bet to make. Incidentally, this is by no means to say that the majority of experts are always right. Far from it. The history of science provides several examples where most authorities in a given field endorsed the wrong theory at one time or another. Most astronomers in Galileo's time thought the Italian scientist got it wrong, but time vindicated Galileo. However, this occasional turn of events cannot be used to justify a systematic rejection of expert opinion. To do so is to engage in what some philosophers informally call the "Van Gogh fallacy." It goes like this: "Van Gogh was an artistic genius, and yet he died penniless. I am a penniless artist, therefore I am a genius." A moment's reflection should convince you that the conclusion simply does not follow from the premise, and you can substitute "Galileo," "scientific genius," and "was shunned by critics" in the phrases above without changing the (logical) outcome at all.

When we move on to Goldman's fourth criterion, we find an apparent obvious mistake, especially for a philosopher. Goldman tells us that one ought to look for evidence of bias, ideological, financial, or of any other kind, that may affect an expert's opinion. His example is of a study published in 1999 in the *Journal of the American Medical Association*, in which the authors surveyed research on anticancer drugs, keeping track of where the funding for the research was coming from. The results were perhaps not surprising to anyone who appreciates human nature, but embarrassing to the medical research community nonetheless: 38 percent of the studies sponsored by nonprofit organizations (such as the federal government) yielded negative results about the efficacy of the tested drugs, while only 5 percent of the studies sponsored by the pharmaceutical industry turned out no effect. If you think the difference is due to chance, I have a beautiful and sturdy bridge in Brooklyn I'd like to sell you . . .

Now, before the postmodernists among us jump up to shout "I told you so," let me point out a couple of things. First, no scientist would really be surprised by the outcome of the study. As we have seen, because science is a human activity, one should expect human biases to enter into it. But, and this is the second point, notice that just like the Piltdown forgery about a fake prehuman ancestor was uncovered by scientists, not creationists, so the 1999 study in *JAMA* was done by scientists, not postmodernists. It is science's ability of self-evaluation, self-criticism, and self-correction that once again shows us both the power and the limits of the scientific enterprise itself. Moreover, the fact of the matter is that it is simply not possible to find any human being entirely without biases. "Bias" is a pejorative term, but more broadly it reflects the fact that we all come with (more or less well founded) opinions and that we can't help but see the world through a series of partial, often distorting, filters. The question is not whether there is bias (there always is), but how much, where it comes from, and how one can become aware of and correct it. That is why I warned the reader about so-called think tanks earlier in the book: too often their political, ideological, and financial biases are not disclosed to the public, which gives them the misleading aura of being neutral, third-party experts.

Following this line of reasoning, however, both Goldman and I are dangerously close to committing the "genetic fallacy" mentioned above, that is, rejecting someone's arguments because of *who* makes the argument rather than on the basis of *why* he makes it. John West, for instance, a fellow of the pro-ID think tank Discovery Institute, has written an entire book, *Darwin Day in America*, complaining about the alleged moral corrosion of modern society brought about by "darwinism." Have you noticed that I biased you against West by mentioning first that he is associated with a pro-creationist think tank? Well, was it bias or fair warning? In making up your mind about West's arguments you should certainly read his book,[29] but knowing that he has a strong ideological background that may predispose him to be antiscience, your baloney detector will be on high alert when you do—as it should be (by the same token, of course, you should consider my own political and philosophical views when reading what I write, and you can find a host of information on this topic on my blog, rationallyspeaking.org).

Nonetheless, neither Goldman nor I are committing any fallacy. The difference is this: if one were to say that we *ought* to reject an argument solely because of the biases of who proposes said argument, then one would in fact be committing the genetic fallacy (one would also be left with few people to argue with, considering that pretty much everyone has one bias or another). But this is not what Goldman and I are saying. All we are proposing is that it is rational (if you believe in rationality) to be aware of the potential effects of a variety of sources of bias, just as the *JAMA* article clearly demonstrated. So the awareness of bias is not an *argument* against someone's position as much as a warning signal that we need to pay close attention to his reasoning and evidence because they may not be as sound or convincing as they might at first glance appear. Just a commonsense form of critical thinking.

Lastly, let's look at the fifth source of evidence available to an intelligent novice attempting to figure out how much trust to put in a self-professed expert: track record. We do this all the time in everyday situations: we keep going to the same mechanic because he has been pretty successful at fixing problems with our car; had he not been, we would have looked for another mechanic. Of course, like every tool described so far, this one needs to be used with caution. A common misuse is what I referred to above as the "Nobel effect," by which people are inclined to accept the pronouncement of a Nobel laureate even if it is in a field remote from the one for which the person in question won the Nobel. The reasoning must go something like this: this person won the Nobel (say, for research in cancer biology); so this person must be smart; therefore, he is probably right about what he just said about, say, economics. This simply does not follow. Someone most certainly must be intelligent (under whatever definition of intelligence we wish to use) to win the Nobel. But intelligence is not the same thing as expertise, and while winning the Nobel in medicine requires expertise in that discipline, it says nothing at all about that person's expertise in a different field, such as economics. Indeed, it suggests that the biologist in question probably is *not* an expert in economics because, as we have seen, it takes a lifetime of effort to be an expert in *either* biology *or* economics, and most of us don't seem to find the equivalent of two lifetimes to devote to disparate professions. On top of this, of course, past success does not guarantee future success: your mechanic may have been infallible so far, and yet he may have finally encountered a problem about your car that he can't solve. He is still an expert, though, and it would be unwise of you to discard his advice, unless he begins consistently to get things wrong.

Goldman is well aware that his five recommendations do not provide us with an airtight method of telling expertise from flimflammery. Indeed, this whole essay is an argument that there is no such foolproof method. But Goldman invites the reader to use what philosophers call "inference to the best explanation,"[30] an approach that seeks to weigh all the available evidence and then make an educated guess about which way the truth lies. This is essentially the same method immortalized by Sir Arthur Conan Doyle in his Sherlock Holmes stories (and which Doyle erroneously referred to as "induction"). When Holmes says, in *Silver Blaze*, "improbable as it is, all other explanations are more improbable still,"[31] he is making an inference to the best explanation compatible with the data as it is available to him. And he is well aware that he needs good information before it becomes possible to arrive at a reasonable inference: in *The Adventure of the Copper Beeches*, Holmes complains, "Data! Data! Data! I can't make bricks without clay."[32] Then again, Holmes's creator himself was not above making major blunders when it came to science and pseudoscience: Conan Doyle famously believed, literally, in fairies, on the grounds that there existed empirical evidence of their appearances in the form of (fake, as it unsurprisingly turned out) photographs.[33]

Judging Expertise: Behe versus Miller, with a Note on Chopra

Let's try to apply all we have learned so far about expertise to a couple of specific examples. I will make an argument that biologist Kenneth Miller of Brown University is more likely to be correct about evolution versus intelligent design than Lehigh biochemist Michael Behe. As a bonus, I will then try to convince you that there is no way that Deepak Chopra can be considered an expert on quantum mysticism, but for a very different reason: there is no such thing as quantum mysticism!

Let us start with Behe versus Miller, both of whom have written books on the topic[34] and both of whom have participated as witnesses (on opposite sides) to the Dover, Pennsylvania, trial that we discussed earlier. On the basis of Goldman's first criterion, how good their arguments are, I have already rendered my verdict. But of course my own conclusion is likely biased, so don't take my word for it. Instead, examine closely the strong legal opinion written by Judge Jones, a conservative appointed by G. W. Bush. As they say in soccer, 1–0 for Miller.

Second round: evidence of agreement by other experts. Although Behe can marshal a few other biologists on his side, the overwhelming majority of professional biologists accepts the theory of evolution and rejects creationism or intelligent design. While we need to remember the caveat that sometimes even a majority of science experts is wrong, the score is now Miller, 2; Behe, 0.

Third: independent evidence that the expert is an expert. Here things are more balanced. Both Behe and Miller have a legitimate Ph.D. in the sciences, and both of them have published peer-reviewed papers in their chosen fields (biochemistry and cell biology, respectively). I'd say this is a tie, the total score goes to 3–1 in favor of Miller.

Fourth: what are their biases? This is particularly interesting, because the two contenders have the same religious background: they are both practicing Catholics. This goes against the oft-repeated creationist mantra that evolutionary biology is a thin veil for a materialist agenda being pushed by a bunch of atheistic scientists. While some scientists are indeed both atheists and vocal opponents of creationism (Richard Dawkins is the obvious example), it is very hard to find an agnostic, atheist, or even a member of a religion besides certain brands of Christianity and Islam on the side of intelligent design.[35] On this count too, then, the two are on the same level; we go to 4–2.

Finally, what about the track record of Behe and Miller? Here Miller, I think, performs better. While it is true that Behe has published peer-reviewed papers, as noted above, none of them has anything to do with evolutionary biology (and his academic productivity has essentially ceased since his creationist career took off). Miller, on the other hand, though a cell biologist, is sufficiently established in the field to have coauthored textbooks that are widely used to teach both general and evolutionary biology. Furthermore, when their specific writings on the evolution-ID controversy are analyzed by scientists and philosophers of science, Behe gets the brunt of the criticisms[36] while Miller is commented upon favorably by a majority of other experts. I'd say this brings us to a final score: Miller, 5; Behe, 2. In soccer terms, this is a rout.

The little exercise above, of course, should not be taken to mean that Ken Miller is definitively right and Michael Behe clearly wrong, though if I had to bet I would certainly put my money on Miller. Nor should this be interpreted as a simplistic reduction of the evolution-creation debate to the antagonism between two of the many contenders. But it is a good way to exemplify how Goldman's five sources of evidence can be used to (provisionally) make up one's mind about a complex matter whose technical aspects one does not understand fully. Try it out yourself: pick recent controversial, opposing claims made by two experts on TV, say, and apply the five steps above, keeping track of your scores and reasons for assigning points. It really makes you think about why you accept (or reject) certain notions.

Now on to Deepak Chopra. He is allegedly an expert in quantum mysticism, and he sells millions of books based on this view of the world. Should we not take him seriously? Not really, and for a variety of reasons, some of which are related to Goldman's criteria, but one of which trumps all other considerations, as far as I can see. First off, it would seem logical to think that "quantum mysticism," whatever that is, has something to do with quantum mechanics. After all, Chopra often uses the jargon of quantum physics in order to fool his audiences and pretend he is actually saying something of substance. But Chopra's (arbitrary) pronouncements are just that, not arguments, thereby failing Goldman's first criterion on the basis that there is *no* argument being made at all! (This, incidentally, is typical of mysticism in general, not just of the Chopra variety.) Second, while real experts in quantum mechanics agree that Chopra just makes things up, other alleged experts in mysticism sharply disagree with Chopra, as attested by the fact that they write their own books claiming to provide *the* answer to life's problems. That answer has, as far as anyone can tell, little in common with Chopra's. Third, Chopra is most definitely *not* an expert in quantum mechanics—or any science, for that matter. Fourth, in terms of biases, Chopra surely has financial ones: he makes millions from selling the nonsense he writes, so one can hardly expect him to be objective when it comes to criticism of mystical "solutions" to people's problems. Lastly, Chopra's track record in terms of accomplishments is dismal: just consider his popular bestseller about how to stave off aging, *Grow Younger, Live Longer: Ten Steps to Reverse Aging*. The funny thing about it is that Deepak's most recent public photos show him significantly aged (if presumably much wealthier) than earlier shots, just like any other mortal who doesn't possess the secrets of quantum mysticism. Could he not apply his own miracle remedies to himself?[37]

It should be clear by now that there are people who can, in fact, be reasonably considered experts; that it is rational to rely, within limits, on expert opinion; and that it is possible, by exercising relatively simple criteria, to gain insight into whether a particular expert is reliable or not. It is also true that experts, of course, do make mistakes, and that even the agreement of a large majority of experts in a field does not guarantee that they got it right. That's the nature of scientific truth, as we have seen throughout this book: it is tentative, because it is the result of a human endeavor that is limited both by the type and amount of available evidence and by humans' finite mental powers and emotional reactions. But the examples above show how you can, with a little bit of practice, tell science from bunk!

CONCLUSION

So, What Is Science after All?

We are now at the end of this book, but hopefully only at the beginning of a long and stimulating intellectual journey. We started out by considering what philosopher Karl Popper called the demarcation problem, the question of how to tell the difference between science and pseudoscience. We immediately rejected Popper's own answer, the simple criterion of falsifiability (that a hypothesis or theory can in principle be refuted based on a single piece of contradictory evidence). Real science is too complicated an activity for such a straightforward solution to apply, and philosophers of science have in fact long since moved beyond falsificationism to more sophisticated and competing views of what distinguishes science from pseudoscience (a discussion of which would quickly lead us into technical territory).[1]

Thus the demarcation problem remains, and throughout this book we examined many of its facets, looking for insights from the history of science, discussions about the alleged difference between soft and hard science, and critical analyses of quasi-science and downright pseudoscience. We inquired into aspects of how science is treated in the media, in politics, and in courts of law. We debated issues like intellectualism and its shortcomings, the science culture wars, and the very nature of expertise, scientific or otherwise. In the end, what picture of science emerges from all of this? How is this picture going to help us be informed citizens in an open society that is increasingly affected—for good or for ill—by science, and that is constantly in danger of being duped by pseudoscience?

For one thing, it should now be clear that science is not a monolithic activity. There are many ways of doing science, depending largely on two factors: the complexity of the object of study (say, subnuclear particles in physics vs. brains in neurobiology) and the degree to which the history of the systems being studied plays a role (none in the case of particle physics, dominating for paleontology or astronomy). The simpler the object of scientific investigation, the more precise our answers are going to be, while the more complex the issues, the more scientists have to rely on statistical approximations. This, as we have seen, has nothing to do with the relative competence of scientists working in different fields, but is a reflection of what our conceptual tools and measurement instruments can tell us about the way nature is. Similarly, the more ahistorical a discipline is, the more successful it will be in making predictions about the behavior of things; but as soon as history plays a role, the emphasis shifts toward a more detective-like work in which scientists are better able to reconstruct what happened rather than predict what might happen.

It appears, therefore, that there is a spectrum between "hard" (i.e., ahistorical, focused on simple subjects) and "soft" (historical and/or dealing with complex subjects) sciences, though of course such distinction is not at all an indication of how "good" the sciences themselves are. What all scientific inquiry has in common, however, are the fundamental aspects of being an investigation of nature, based on the construction of empirically verifiable theories and hypotheses. These three elements, *naturalism*, *theory*, and *empiricism*, are what make science different from any other human activity. Since science deals with natural phenomena and processes, it by definition has no business getting into supernaturalism, be it in the form of misguided attempts by scientists to "prove" the existence (or inexistence) of gods, or in the form of notions that are based on supernaturalism like intelligent design creationism. (Of course, I do not mean to say that scientists should not criticize creationism; rather, I mean to suggest that if the claim is about supernatural powers, it is not about science.)

The presence of coherent conceptual constructs in the form of theories and hypotheses is also a necessary component of science. Science is not just a collection of facts about the world, nor do scientific theories emerge from the accumulation of facts, as Francis Bacon thought. Theories are creative productions of the human mind and reflect our best attempts at making sense of the world as it is. But theories are not enough, otherwise science would be no different from philosophy. It is the crucial role of empirical information that completes the trinity that underlies all scientific research. Empirical evidence, as we have seen in this book, does not necessarily mean experiment, but more broadly refers to any combination of experimentation and systematic observation that produces not just facts, but data.

Empirical testability, then, is one major characteristic distinguishing science from nonscience. Although something might sound "scientific," such as in the case of string theory in physics or the borderline examples of evolutionary psychology and the search for extraterrestrial intelligence, a field does not belong to science unless there are reasonable ways to test its theories against data. Plenty of human activities, of course, are not scientific in this sense. Personal taste in, say, art may be subject to empirical surveys (we can ask people what they like and dislike), and taste clearly is an aspect of nature, since human culture is as natural as anything else. But unless our research on taste is informed by an overall conceptual structure (a theory) that can be used to generate specific testable hypotheses, it is not science.

So what about pseudoscience? What picture emerges there? A field can fall into that category because it again fails to meet the three criteria of naturalism, theory, and empiricism, and yet its supporters insist that there is no problem. Creationism, most obviously, does not satisfy the first criterion. We have seen that astrology is based on an embryonic theory of sorts, is subject to empirical investigation, and—at least in most forms—does not invoke supernatural influences. What makes astrology a pseudoscience, then, is that its practitioners keep defending it despite the fact that its theoretical structure is hopelessly flawed (e.g., constellations do not exist) and that when empirically tested it repeatedly fails the confrontation with the data. We have even seen that initially sound scientific inquiries, like the one on the possibility of cold fusion, may slip into the dark territory of pseudoscience when people refuse to bow to the testability criterion. All of which means that the boundaries separating science, nonscience, and pseudoscience are much fuzzier and more permeable than Popper (or, for that matter, most scientists) would have us believe. There is, in other words, no litmus test for whether something is good science or not. It depends on a variety of factors, and the role played by these factors may change over time so that a previously fringe discipline may blossom into mature science or, conversely, a formerly respectable field may acquire the characteristics of pseudoscience.

The other conclusion that should be obvious by now is that science itself, even good science, is no guarantor of Truth (notice the capital *T*). Science is a complex social activity carried out by limited human beings who are affected by the time and place in which they happen to live, not to mention by having a brain that evolved to solve everyday life problems, not to rationally and impartially pursue cosmic questions about the nature of things. For every amusing prank played by scientists like Alan Sokal at the expense of naïve humanists and assorted nonscientists, it is easy enough to trot out a long litany of blunders that self-assured scientists have committed over the centuries, or even very recently. That, in and of itself, ought to be sufficient warning not to trust science (and even less individual scientists) without further questioning.

It is here that we reach the limit of our investigation into the nature of science. We have pretty good—though by no means infallible—tools to tell us what's bad science or pseudoscience, and we can identify easily enough most of what counts for solid science. But when it comes to betting on the reliability of individual scientific statements or assessing the real status of scientific disputes such as the one about global warming, there is no simple solution. Expertise in the matter is an indication that someone probably knows better than others what he or she is talking about. But we have also seen that experts can be spectacularly wrong and that it is difficult to ascertain someone's level of expertise or ideological bias. A "Ph.D." after an author's name on a book cover is by no means a reliable indication of the soundness of what one may find inside the book.

Ultimately, there are no shortcuts; we must seriously engage the issues at hand, in proportion to how much we care about them and to our responsibility to society. If you are concerned about the environment, regardless of your other political positions, you may need to spend time reading various opinions about pertinent environmental issues, making up your mind the best you can. If your vote at the ballot box depends on issues in which science has something to say, then by all means listen to what scientists are saying about it.

But never, ever forget to turn on your baloney detector. Most of the time you will need it set at least to yellow alert.

Notes

Chapter One

1. J. R. Platt, "Strong Inference," *Science* 146 (1964): 347–53.

2. For a somewhat technical treatment of this see: M. Pigliucci, "Do We Need an Extended Evolutionary Synthesis?" *Evolution* 61, no. 12 (2007): 2743–49.

3. Platt, "Strong Inference," 347.

4. Yes, I am aware of the existence of novel trends in twentieth-century symphonic music. I am simply arguing that they have been much less spectacular than their counterparts in jazz—just as not even Platt would argue that the "soft" sciences do not make any progress at all, only that they are comparatively much slower than the hard ones.

5. Not to mention other "-omics," such as proteomics (the study of the properties of all proteins produced within a cell), metabolomics (all chemical products of metabolism), and even phenomics (the study of pretty much everything that happens in a biological organism).

6. In Larry V. Hedges, "How Hard Is Hard Science, How Soft Is Soft Science?" *American Psychologist* 42, no. 2 (1987): 443–55.

7. George S. Howard, "When Psychology Looks like a 'Soft' Science, It's for Good Reason!" *Theoretical and Philosophical Psychology* 13, no. 1 (1993): 42–47.

8. However, see M. Pigliucci, "Can There Be a Science of Free Will?" *Skeptical Inquirer* 31, no. 3 (2007): 26–27.

9. H. Gee, *In Search of Deep Time* (New York: The Free Press, 1999), 267.

10. C. E. Cleland, "Historical Science, Experimental Science, and the Scientific Method," *Geology* 29 (2001): 987–90. For a more technical presentation of the same material, see also C. E. Cleland, "Methodological and Epistemic Differences between Historical Science and Experimental Science," *Philosophy of Science* 69 (2002): 474–96.

Chapter Two

1. Steven Weinberg, "Against Philosophy," in *Dreams of a Final Theory: The Scientist's Search for the Ultimate Laws of Nature* (Vintage, 1994).

2. "Physics Ain't What It Used to Be," *Nature* 438 (2005): 739–40.

3. "Pulling the Strings" *Nature* 438 (2005): 1081–82.

4. For instance, see Benjamin Wallace-Wells, "Surfing the Universe," *New Yorker*, 21 July 2008.

5. Brian Greene, *The Elegant Universe: Superstrings, Hidden Dimensions, and the Quest for the Ultimate Theory* (W. W. Norton & Co., 2003).

6. See, for instance, R. De Sousa, "Is Art an Adaptation? Prospects for an Evolutionary Perspective on Beauty," *Journal of Aesthetics and Art Criticism* 62, no. 2 (2004): 109–18; as well as J. D. Greene, R. B. Somerville, L. E. Nystrom, J. M. Darley, and J. D. Cohen, "An fMRI Investigation of Emotional Engagement in Moral Judgment," *Science* 293 (2001): 2105–8.

7. Available at http://setiathome.berkeley.edu/.

8. Available at http://www.classbrain.com/artmovies/publish/article_50.shtml.

9. We also included a series of messages recorded on a golden disk on the *Voyager* spacecraft back in 1977 (incidentally, some of the promoters of this second enterprise were the same ones that were behind the Arecibo message, including astronomer Carl Sagan). The *Voyager* would take about forty thousand years to reach the nearby star Alpha Centauri, except that it is not traveling in the right direction . . .

10. Edward O. Wilson, *Sociobiology: The New Synthesis* (Belknap Press, 1975).

11. Edward O. Wilson, *On Human Nature* (Harvard University Press, 1978).

12. Richard J. Herrnstein and Charles Murray, *The Bell Curve: Intelligence and Class Structure in American Life* (Free Press, 1994).

13. Randy Thornhill and Craig T. Palmer, *A Natural History of Rape: Biological Bases of Sexual Coercion* (MIT Press, 2001).

14. Richard Lewontin, "The Evolution of Cognition: Questions We Will Never Answer," in *An Invitation to Cognitive Science*, ed. D. N. Osherson (MIT Press, 1998), 4:107–32.

15. John L. Casti, *Paradigms Lost* (Harper Perennial, 1990).

16. The classic paper on the swordtail fish is A. L. Basolo, "Female Preference Predates the Evolution of the Sword in Swordtail Fish," *Science* 250 (1990): 808–10. A more recent work on the predation aspect of the story is in G. G. Rosenthal, T. Y. Flores-Martinez, F. J. Garcia-de-Leon, and M. J. Ryan, "Shared Preferences by Predators for Male Ornaments in Swordtails," *American Naturalist* 158 (2001): 146–54.

17. I do not actually know whether Napoleon had diarrhea that fateful morning, but that is precisely the point.

18. See J. Goodheart, *Darwinian Misadventures in the Humanities* (Transaction Publishers, 2007), 67.

19. Ibid., 71.

20. Ibid., 76.

21. "Arise 'Cliodynamics,'" *Nature*, 3 July 2008.

Chapter Three

1. Published on 13 April 2008.

2. An even more recent case happened in Papua New Guinea while I was revising the manuscript for this book, though this time the motive had to do with the mysterious death of a neighbor and possibly with superstition concerning the spread of HIV/AIDS. See: http://rationallyspeaking.blogspot.com/2009/01/superstition-can-kill-you.html.

3. Including, but most certainly not limited to: James Randi, *Flim-Flam! Psychics, ESP, Unicorns, and Other Delusions* (Prometheus, 1982); C. Sagan, *The Demon-Haunted World: Science as a Candle in the Dark* (Random House, 1995); N. Humphrey, *Leaps of Faith* (Basic Books, 1996); M. Gardner, *Did Adam and Eve Have Navels? Debunking Pseudoscience* (Norton, 2001); M. Shermer, ed., *The Skeptic Encyclopedia of Pseudoscience* (ABC Clio, 2002). You will have noticed that some of these books are a bit old—don't worry, one of the major differences between real science and pseudoscience is that the latter hardly changes at all. It is also worth checking out the popular ShowTime TV series *Bullshit!* hosted by magicians Penn and Teller. However, watch out for some (rather gratuitous, in my opinion) strong language and a sometimes overt libertarian political bias.

4. "Aids Victims Risk Lives," 3 April 2007.

5. "The Denialists: The Dangerous Attacks on the Consensus about H.I.V. and AIDS," 12 March 2007.

6. An even more outrageous case of Big Pharma caught cheating was reported in March 2009. The Merck company, which produced the drug Vioxx, linked to thousands of cases of heart attack and stroke, set up a fake journal (with the complicity of otherwise respectable publisher Elsevier) so that they could bypass the standard scientific process of peer review and publish papers favorable to their own products. Fortunately, a lawsuit has exposed the practice, which Elsevier has ceased. (See http://abcnews.go.com/Health/Drugs/story?id=7577646&page=1.)

7. A nicely written and accessible article on this is again by Michael Specter: "Darwin's Surprise: Why are Evolutionary Biologists Bringing Back Extinct Deadly Viruses?" *New Yorker*, 3 December 2007.

8. See P. Plait, *Bad Astronomy: Misconceptions and Misuses Revealed, from Astrology to the Moon Landing "Hoax"* (Wiley, 2002).

9. Geoffrey O. Dean and Ivan W. Kelly, "Is Astrology Relevant to Consciousness and Psi?" *Journal of Consciousness Studies* 10, no. 6–7 (2003): 175–98.

10. If you want an idea of just how negligible, it suffices to say that the next largest influence on terrestrial tides (after the moon and sun) is exerted by Venus: it is 0.000113 times that of the sun.

11. That is just the order of magnitude of the known asteroids, but the estimate is of 750,000 asteroids larger than one kilometer in diameter, and of millions of smaller ones.

12. See Dean and Kelly, "Is Astrology Relevant to Consciousness and Psi?" cited above.

13. David Voas, "Ten Million Marriages: An Astrological Detective Story," *Skeptical Inquirer*, March/April 2008.

14. S. Carlson, "A Double-Blind Test of Astrology," *Nature* 318 (1985): 419–25.

15. See the commentary by Paul Kurtz in *Skeptical Inquirer*, July/August 1997 issue.

16. Most of the available literature is, predictably enough, pro-ufology. Nonetheless, some noteworthy entries on the side of critical thinking include: Philip Klass, *Ufos: The Public Deceived* (Prometheus, 1986); Curtis Peebles, *Watch the Skies!* (Berkley, 1995); Kendrick Frazier, Barry Karr, and Joe Nickell, eds., *The UFO Invasion* (Prometheus, 1997); Robert Sheaffer, *UFO Sightings: The Evidence* (Prometheus, 1998); Susan Clancy, *Abducted: How People Come to Believe They Were Kidnapped by Aliens* (Harvard University Press, 2007).

17. James Oberg, "The Great East Coast UFO of August 1986," *Skeptical Inquirer*, Winter 1986–87.

18. For a report on the investigation of the cold case, see Joe Nickell, "Investigative Files: the Flatwood UFO Monster," in *Skeptical Inquirer*, November 2000.

19. For a quite technical, but engaging, discussion of how to assess scientific evidence, see chapter 1 of E. Sober, *Evidence and Evolution: The Logic behind the Science* (Cambridge University Press, 2008).

20. See Robert Sheaffer, "The Campeche, Mexico, 'Infrared UFO' Video," *Skeptical Inquirer*, September 2004.

21. See James Oberg, "The Failure of the 'Science' of UFOlogy," *New Scientist*, 11 October 1979.

22. Indeed, the latest example of such a tactic appeared while I was writing this chapter: according to the *New York Times* (24 July 2008), former astronaut Edgar Mitchell, the sixth man to walk on the moon, has said that he is "privileged enough to be in on the fact that we've been visited on this planet and the U.F.O. phenomena [*sic*] is real." Mitchell has made this sort of declaration before, but offered no supporting evidence. Moreover, his Noetic Science Institute is well known for attracting believers in extraterrestrial visitations, but not known at all for the production of quality scientific work.

23. Again, the list would be endless, but some notable entries include: the already mentioned James Randi, *Flim-Flam! Psychics, ESP, Unicorns, and Other Delusions*; Paul Kurtz, ed., *Skeptical Odysseys: Personal Accounts by the World's Leading Paranormal Inquirers* (Prometheus, 2001); Terence Hines, *Pseudoscience and the Paranormal* (Prometheus, 2003); Massimo Polidoro, *Secrets of the Psychics: Investigating Paranormal Claims* (Prometheus, 2003); Lynne Kelly, *The Skeptic's Guide to the Paranormal* (Basic Books, 2005); Joe Nickell, *Adventures in Paranormal Investigation* (University Press of Kentucky, 2007).

24. Lucy Odling-Smee, "The Lab That Asked the Wrong Questions," *Nature* 1 March 2007.

25. Stanley Jeffers, "The PEAR Proposition: Fact or Fallacy?" *Skeptical Inquirer*, May/June 2006.

26. Indeed, one of my favorite Holmes quotes is the following, from *The Sign of the Four* (1890): "How often have I said to you [Watson] that when you have eliminated the impossible, whatever remains, however improbable, must be the truth?"

Chapter Four

1. Just for fun, do a vanity search on Google for my full name. The last time I tried this, I found a whopping 30,100 results, an astounding number for someone who is certainly not a celebrity. Now try a "real" celebrity, Britney Spears: I got 69.4 million results . . .

2. Lou Robson, "Witches Brave the Fires of Scepticism," *Age*, 3 July 2008, available online at http://www.theage.com.au.

3. See http://www.talkorigins.org/faqs/icr-visit/bartelt4.html.

4. The group's Web site is: http://www.senseaboutscience.org.uk/.

5. Robert Sheaffer, "Massive Uncritical Publicity for Supposed 'Independent UFO Investigation' Demonstrates Media Gullibility Once Again," *Skeptical Inquirer*, September 1998.

6. Philip J. Klass, "That's Entertainment! TV's UFO Coverup," *Skeptical Inquirer*, November 1996.

7. For a segment-by-segment critique of this episode of *Bullshit!* complete with references and links to back up the criticisms, see: http://www.logicalscience.com/skeptics/bullshit.html.

8. For a detailed discussion of Stewart on evolution, see M. Pigliucci, "Evolution, Schmevolution: Jon Stewart and the Culture Wars," in *The Daily Show and Philosophy: Moments of Zen in the Art of Fake News* (Blackwell, 2007), 190–202.

9. Eric Scerri, "What the #$*! do they know?" *Skeptical Inquirer*, September 2004.

10. There is an alternative interpretation of the relation between the microscopic and the macroscopic worlds in which the wave function does not collapse, but rather is infinitely split (or "dechoered") to form an endless series of parallel universes. This second interpretation is apparently favored by quantum physicists (and *Star Trek* fans!) nowadays, but it doesn't alter my argument about the inanity of *What the Bleep*.

11. As I was writing the first draft of this chapter, in the summer of 2008, a similar splash was being made by yet another self-styled documentary that is nothing of the kind: Ben Stein's *Expelled!* on the alleged anti-intelligent design conspiracy that is wrecking—according to him—our academic system. You can read Michael Shermer's review of the movie in *Scientific American* online at http://www.sciam.com/article.cfm?id=ben-steins-expelled-review-michael-shermer.

12. The proceedings volume will be published by MIT Press in 2010.

13. E. Pennisi, "Modernizing the Modern Synthesis," *Science*, 11 July 2008, 196–97. See also J. Whitfield, "Postmodern Evolution?" *Nature*, 18 September 2008, 281–84.

Chapter Five

1. In A. Etzioni and A. Bowditch, eds., *Public Intellectuals: an Endangered Species?* (Rowman & Littlefield, 2006), 2.

2. Ibid.

3. Ibid., 3.

4. David Carter, "Public Intellectuals, Book Culture and Civil Society," *Australian Humanities Review*, December 2001.

5. Quoted by Frank Furedi in *Where Have All the Intellectuals Gone?* (Continuum, 2004), 7.

6. Noam Chomsky, "The Responsibility of Intellectuals," *New York Review of Books*, 23 February 1963.

7. Ibid.

8. Etzioni and Bowditch, *Public Intellectuals*, 12.

9. In his *Unweaving the Rainbow: Science, Delusion, and the Appetite for Wonder* (Mariner Books, 1998).

10. Frank Furedi, *Where Have All the Intellectuals Gone?* (Continuum, 2004), 2.

11. An in-depth discussion and update of Hofstadter's classic can be found in Daniel Rigney's "Three Kinds of Anti-Intellectualism: Rethinking Hofstadter," published in *Sociological Inquiry* in 1991.

12. This remark most definitely does not imply that I think UFOs and SETI are on the same level, but only that, as David Hume realized back in the eighteenth

century, even the most logical pursuit requires the motivation that comes from passion, or it isn't a pursuit at all.

13. Michael Shermer, *The Borderlands of Science: Where Sense Meets Nonsense*. (Oxford University Press, 2002), chapter 10.

14. All quotations are from Shermer's *Borderlands of Science*, chapter 10.

15. Robert Wright, "The Accidental Creationist," *New Yorker*, 13 December 1999.

16. Ibid.

17. F. De Waal, *Primates and Philosophers: How Morality Evolved* (Princeton University Press, 2006).

18. D. Stone, "Think Tanks, Policy Advise and Governance," in *Think Tank Traditions: Policy Research and the Politics of Ideas*, ed. D. Stone and A. Denham (Machester University Press, 2004).

19. Donald Abelson, "The Business of Ideas: The Think Tank Industry in the USA," in *Think Tank Traditions*, 215–29.

20. Abelson, "Business of Ideas," 220.

21. Leila Hudson, "The New Ivory Towers: Think Tanks, Strategic Studies and 'Counterrealism,'" *Middle East Policy*, Winter 2005.

22. Thomas Osborne, "On Mediators: Intellectuals and the Ideas Trade in the Knowledge Society," *Economy and Society*, 4 November 2004.

23. Alina Gildiner, "Politics Dressed as Science: Two Think Tanks on Environmental Regulation and Health," *Journal of Health, Politics, and Law*, April 2004.

24. Ibid., 317.

25. "About CEI," at cei.org.

Chapter Six

1. Bjørn Lomborg, *The Skeptical Environmentalist* (Cambridge University Press, 2001), 259.

2. Ibid., 263.

3. Ibid., fig. 134.

4. Ibid., 265.

5. Ibid.

6. Ibid.

7. Ibid., 267.

8. Ibid., 272.

9. Ibid., 274.

10. Ibid.

11. Ibid., 277.

12. Ibid., 278.

13. Ibid., 284.

14. Ibid., 289.

15. Ibid.

16. Ibid., 290.

17. Ibid.

18. Ibid.

19. Ibid., 291.

20. Janet Larsen, "Record Heat Wave in Europe Takes 35,000 Lives," available at the Earth Policy Institute Web site: http://www.earth-policy.org/Updates/Update29.htm.

21. Lomborg, *Skeptical Environmentalist*, 291.

22. Ibid., 298.

23. Ibid., 299.

24. Ibid., 319.

25. Ibid., 321.

26. Ibid., 324.

27. See Eban Goodstein, "Hot Air," *Salon*, 29 August 2007.

28. Ibid.

29. I hope the reader will appreciate how easily I accumulated endnotes in this chapter: twenty-eight just to comment on one chapter of Lomborg's book, though I am certainly not in his league when it comes to note-based "scholarship."

30. Katharine Mieszkowski, "Did Al Get the Science Right?" *Chicago Sun-Times*, 18 June 2006.

31. Al Gore, *An Inconvenient Truth* (Rodale Books, 2006), 66–67.

32. Ibid., 67.

33. Ibid.

34. A. O. Scott, "Warning of Calamities with a Scholarly Tone," *New York Times*, 24 May 2006.

35. Mieszkowski, "Did Al Get the Science Right?"

36. Matt Ridley, "The Greens Have Got It Wrong," *Sunday Telegraph*, 26 August 2001.

37. Andrew Greeley, "Doomsday Prophecies Lack Merit," *Chicago Sun-Times*, 9 December 2001.

38. Jonathan Adler, "Dissident from Denmark," *National Review*, 8 April 8 2002.

39. Charles T. Rubin, "Green No More: The Education of an Environmentalist," *Weekly Standard*, 24 December 2001.

40. Stephen Schneider, "Global Warming: Neglecting the Complexities," *Scientific American*, January 2002 issue.

41. Michael Grubb, "Relying on Manna from Heaven?" *Science*, 9 November 2001.

42. Stuart Pimm and Jeff Harvey, "No Need to Worry about the Future," *Nature*, 8 November 2001.

43. Simon Grose, "The Inconvenient Realities behind an Inconvenient Truth," *Canberra Times*, 2 November, 2006.

44. Andrew C. Revkin, " 'An Inconvenient Truth': Al Gore's Fight against Global Warming," *New York Times*, 22 May 2006.

45. Mieszkowski, "Did Al Get the Science Right?"

46. David Remnick, "Ozone Man," *New Yorker*, 24 April 2006.

47. Joe Garofoli, "Gore Movie Reaching the Red States, Too," *San Francisco Chronicle*, 8 July 2006.

48. IPCC 2007 Report, 2.

49. Ibid., 3.

50. Ibid., 10.

51. Ibid.

52. IPCC 2007 Report, 12.

53. Ian Sample, "Scientists Offered Cash to Dispute Climate Study," *Guardian*, 2 February 2007.

Chapter Seven

1. For a fuller treatment of the various and complex aspects of the evolution-creation wars, see Massimo Pigliucci, *Denying Evolution: Creation, Scientism, and the Nature of Science* (Sinauer, 2002).

2. There are several books on the Dover trial that will be most useful to the reader interested in a deeper understanding of what went on and why: Gordy Slack, *The Battle Over the Meaning of Everything: Evolution, Intelligent Design, and a School Board in Dover, PA* (Jossey-Bass, 2007); Matthew Chapman, *40 Days and 40 Nights: Darwin, Intelligent Design, God, OxyContin® and Other Oddities on Trial in Pennsylvania* (HarperCollins, 2007); Edward Humes, *Monkey Girl: Evolution, Education, Religion, and the Battle for America's Soul* (Ecco, 2007); Lauri Lebo, *The Devil in Dover: An Insider's Story of Dogma v. Darwin in Small-Town America* (New Press, 2008).

3. *Tammy Kitzmiller et al. v. Dover Area School District*, case n. 04cv2688, memorandum opinion (20 December 2005). Henceforth *Kitzmiller v. Dover*.

4. *Kitzmiller v. Dover*, 99.

5. Ibid., 100.

6. Ibid., 104.

7. Ibid.

8. *Kitzmiller v. Dover*, 108.

9. Ibid., 110.

10. Ibid., 111.

11. Ibid., 114.

12. Ibid., 123.

13. Ibid., 2.

14. There is a more intriguing variant of the Mount Rushmore argument: what if we found the faces on Mars instead of Earth? ID proponents say that we would still recognize them as the result of intelligent design even though that design couldn't possibility be human. This is indeed the case, but notice that we would infer a humanlike designer, not a supernatural one. Why? Because, again, we would make use of our existing knowledge about intelligent design, and it is that knowledge—not just the observations of peculiar-looking structures per se—that licenses the "design inference."

15. *Kitzmiller v. Dover*, 124.

16. Ibid., 129.

17. Ibid., 125.

18. Ibid., 128.

19. Ibid., 58.

20. Ibid., 131.

21. E. J. Larson, *Summer for the Gods. The Scopes Trial and America's Continuing Debate over Science and Religion* (Basic Books, 1997).

22. B. Forrest and P. R. Gross, *Creationism's Trojan Horse: The Wedge of Intelligent Design* (Oxford University Press, 2004).

23. *Kitzmiller v. Dover*, 9.

24. In *Lynch v. Donnelly*, 465 U.S. 668, 688 (1984).

25. *Kitzmiller v. Dover*, 90.

26. Ibid., 64.

27. Ibid.

28. O'Connor, E. 1993. The Essential Epicurus. Prometheus Books, Amherst, NY, p. 49.

29. Of course, one can always make a claim that one personally knows God's motives. That tactic, however, is not only unscientific (and quite a bit egomaniacal), it can easily be countered by another person's assertion of the same divine privilege, but leading to a different conclusion. How are we to know to whom God really speaks?

30. *Kitzmiller v. Dover*, 30.

31. Ibid., 68.

32. Behe more recently published a second book on irreducible complexity, *The Edge of Evolution: The Search for the Limits of Darwinism* (Free Press, 2007), but he did not advance his argument beyond the original version.

33. W. Paley, *Natural Theology; or, Evidences of the Existence and Attributes of the Deity, Collected from the Appearances of Nature* (Gould, Kendall, and Lincoln, 1802).

34. W. A. Dembski, *The Design Inference* (Cambridge University Press, 1998). For a critique of Dembski's book, see B. Fitelson, C. Stephens, and E. Sober, "How Not to Detect Design—Critical Notice: William A. Dembski, *The Design Inference*," *Philosophy of Science* 66 (1999): 472–88; M. Pigliucci, "Chance, Necessity, and the Holy War against Science: A Review of W. M. Demski's *The Design Inference*," BioScience 50 (2000): 79–87.

35. There is another reason why ID supporters tend to be vague about the designer's attributes: theirs is a political movement, and the intention is to set up as broad a tent as possible, to include creationists of all stripes. If one begins to specify attributes of the designer, one risks alienating part of that politically necessary (and often theologically cantankerous) base.

36. M. J. Behe, *Darwin's Black Box: The Biochemical Challenge to Evolution* (Free Press, 1996). For a critique, see N. Shanks and K. H. Joplin, "Redundant Complexity: A Critical Analysis of Intelligent Design in Biochemistry," *Philosophy of Science* 66 (1999): 268–82; K. R. Miller, *Finding Darwin's God: A Scientist's Search for Common Ground between God and Religion* (Harper Collins, 1999).

37. Available at http://udel.edu/~mcdonald/mousetrap.html.

38. See D.-E. Nilsson and S. Pelger, "A Pessimistic Estimate of the Time Required for an Eye to Evolve," *Proceedings of the Royal Society London* B 256 (1994): 53–58.

39. F. Jacob, "Evolution and Tinkering," *Science* 196 (1977): 1161–66.

40. See Shanks and Joplin, "Redundant Complexity."

41. *Kitzmiller v. Dover*, 76.

42. Ibid., 77.

43. Ibid., 78.

44. Ibid., 81.

Chapter Eight

1. For a rather detailed but enjoyable history of early science through the Middle Ages, see D. C. Lindberg, *The Beginnings of Western Science* (University of Chicago Press, 1992).

2. The term "pre-Socratic" may appear a bit funny, as if Socrates were a Christ-like figure that divides the history of philosophy into what happened before him and after him. Actually, some people do see similarities between the two men, but the reasons for drawing a line between philosophy before and after Socrates don't have much to do with the mystique of the Athenian thinker. For one thing, we have only mere fragments of writings attesting to what the pre-Socratics thought, while we have a detailed account of the teaching of Socrates (ironically, none of it written by him, but rather by his disciple, Plato). For another thing, the type of philosophy started by Socrates, Plato, and Aristotle has, essentially, set the agenda

for the following two millennia of discussions. Finally, Socrates was put to death in 399 BCE, thereby providing philosophers with their first official martyr. In that respect, of course, his figure does have parallels with that of Jesus.

3. For more on string theory, see its "official" Web site at http://www.superstringtheory.com. I'm not sure what makes it "official," but there it is.

4. This description does bear a caveat, however: the "empty" space is permeated by energy fields, but energy, according to modern physics, is simply another state of matter (and vice versa); so, strictly speaking, the empty space isn't empty after all. Still, all of this has, obviously, nothing to do with our commonsensical perception of objects, which was my point to begin with.

5. As I mentioned, however, Socrates left nothing in writing, and most of what we know about his thinking is filtered through Plato. The latter is in fact so crucial to the history of philosophy that Alfred North Whitehead famously quipped that all philosophy is but a footnote to Plato. An exaggeration, no doubt, but a nod to the fact that Plato systematically began discussions on pretty much all of the major topics that philosophers care about. This is certainly not a small accomplishment, though perhaps made easier by the fact that there wasn't much history of philosophical discourse before him.

6. Should you be interested in that story, you will find out more here: E. Sober and D. S. Wilson, *Unto Others: The Evolution and Psychology of Unselfish Behavior* (Harvard University Press, 1998); F. De Waal, *Primates and Philosophers: How Morality Evolved* (Princeton University Press, 2006); and M. Pigliucci, "Is Ethics a Science?" *Philosophy Now*, May/June 2006, 25.

7. For more on rationalists versus empiricists, see R. S. Woolhouse, *The Empiricists* (Oxford University Press, 1988); and J. Cottingham, *Rationalism* (Thoemmes Press, 1997).

8. In case you are wondering, this is the very same Berkeley after whom the famous University of California campus and surrounding town are named.

9. Locke's tabula rasa idea is the predecessor of the modern "nurture" position (as opposed to "nature") in the never-ending debate on the roles of genes and environments in shaping human cognitive traits. The current scientific consensus is that human traits, like those of other animals, plants, and other organisms, are the result of complex interactions between genes and environment. For an accessible treatment of this matter, see M. Ridley, *Nature via Nurture: Genes, Experience, and What Makes Us Human* (HarperCollins, 2003). For a more technical version, see M. Pigliucci, *Phenotypic Plasticity: Beyond Nature and Nurture* (Johns Hopkins University Press, 2001).

10. Kant's life in general, apparently, was not that exciting: legend has it that he took his daily walks so regularly that the good citizens of Konigsberg would set their clocks by his passage. This is probably apocryphal, but it gives us an idea of the punctuality (and repetitiveness) of Kant's routine.

11. I. Kant, *Prolegomena to Any Future Metaphysics* (1783), 8.

12. Incidentally, one of Kant's insights was precisely that we can never know the world as it is "in itself," because our grasp of it will always be mediated by the Ideas. This conclusion would also seem to undermine the mystical claims of Eastern religions.

13. The phrase "platonic love" originally referred to this sense of "ideal" or "perfect." Plato never said that love should be nonphysical.

14. Again, this is not to say that science, or philosophy for that matter, is necessarily atheistic, because of the distinction between philosophical and methodological naturalism introduced in the previous chapter.

15. He called it "first philosophy." The term "metaphysics" originally simply meant the subject matter treated in the books Aristotle wrote after the *Physics* and was introduced by some of the early editors of Aristotle's work.

16. Yes, this is the very same island of Lesbos from which the word "lesbian" comes. It, however, had nothing to do with Aristotle's biological research. Instead, that meaning is associated with a poem by Sappho, who was born on Lesbos, and which is in fact about strong feelings of love between women.

17. Incidentally, Aristotle is often dismissed in the context of discussions of the scientific method on the ground that he limited himself to observation without engaging in experiments—the hallmark of modern science. But as Lindberg points out (*Beginnings of Western Science*, 53), Aristotle made a choice based on his understanding of science, not out of ignorance. For Aristotle, natural objects had to be investigated in their natural context, because doing otherwise would have yielded artificial results of unclear applicability. This position is still maintained by a number of scientists (especially in ecology) today, and it certainly does not represent a limitation of Aristotle's thought but rather a legitimate epistemological position.

18. Philosophers sometimes refer to this position as the "no ectoplasm clause," meaning that however the brain causes the mind, one thing we know for sure: there are no ghosts (ectoplasms), because whenever the brain stops functioning properly so does mental activity.

19. For an in-depth discussion of the role of Aristotelian causes in biology, see M. Pigliucci, "Are Ecology and Evolutionary Biology 'Soft' Sciences?" *Annales Zoologici Finnici* 39 (2002): 87–98.

20. Lindberg, *Beginnings of Western Science*, 137.

21. However, an argument can be made that the emergence of science after the Middle Ages proceeded in some ways from the inability of the scholastics to solve specific questions within their established framework, which motivated the formulation of new models. In this indirect way, then, Scholasticism becomes part of the history of science.

22. Lindberg, who otherwise strikes a middle-ground position on these controversies in *The Beginnings of Western Science*, claims that to "recount Muslim contributions to the various sciences would require volumes" (175) and yet is unable

to cite much in the way of examples, except for disputes among Islamic scholars on detailed aspects of the Ptolemaic description of the solar system.

Chapter Nine

1. As I already mentioned, for a more in-depth treatment of the history of science up to the Renaissance, see Lindberg's excellent *The Beginnings of Western Science*. A more recent book on the relation between the Middle Ages and the birth of science is Philip Ball's *Universe of Stone: Chartres Cathedral and the Triumph of the Medieval Mind* (Bodley Head, 2008).
2. Lindberg, *Beginnings of Modern Science*, 211.
3. Ibid., 217.
4. Ibid., 226.
5. This debate, like many others, cannot easily be characterized as two-sided: while Aquinas's contribution turned out, in retrospect, to favor an independent development of science, it is clear that Aquinas himself did not intend to subvert the role of theology in intellectual life. Equally clearly, Galileo's status as a pivotal figure in modern science does not negate that he was a devout Catholic who was seriously troubled by his long struggle with the pope.
6. *New Organon*, book 1, *Aphorisms*, 39–68.
7. Notice that an argument can be made that math is not actually a science but, rather, a branch of logic (which itself—ironically—can be thought of as a branch of philosophy). Of course mathematical tools are commonly used in science, but the idea is that for a field of inquiry to qualify as science, it has to have something to do with empirical evidence. Mathematics does not even strictly require the existence of an external world to operate and therefore is in no way directly concerned with it.
8. Of course, these subquestions are in fact still too big, and individual scientific papers are aimed at even smaller chunks of the problem.
9. Indeed, one of the most popular databases of scientific literature is aptly called "The Web of Science."
10. Galileo gave his friends various reasons for moving back to Florence, including that the wine in Padua wasn't that good and that he had too many students to teach. Most modern academics would sympathize with both complaints.
11. This, of course, sounds uncannily like the modern creationist strategy of declaring evolution "just a theory." It is to the chagrin of some contemporary creationists that their movement no longer has the power of enforcement that was available to the church of Rome in the seventeenth century.
12. Contrary to popular belief, Galileo did not invent the telescope, but he greatly enhanced the original version that had been invented a few years earlier

in the Netherlands and made it possible for it to be used as a scientific instrument. Another legend concerning Galileo is that, after having been condemned by the church, he struck the ground beneath him and muttered "Eppur si muove!" ("And yet it moves!"). This is probably a fictional embellishment of a later commentator, Giuseppe Baretti, who wrote about Galileo in 1757. The phrase is found in Berthold Brecht's famous play on Galileo's life.

13. It is rather ironic that thought experiments nowadays are regarded as a poor substitute for real science, mostly indulged in by philosophers. Galileo and, in fact, Einstein are responsible for some of the most beautiful and effective contributions to science, accomplished by means of thought experiments. See M. Pigliucci, "What Is a Thought Experiment, Anyhow?" *Philosophy Now* 58 (2006): 30.

14. The conditions for the experiment were ideal on the moon because there is no friction caused by the atmosphere to interfere with the predictions based on Galileo's theory of motion.

15. Although, as was the case for Galileo mentioned above, we need to distinguish myth from reality: the famous story of Newton suddenly struck with the idea of gravity when his head was physically hit by a falling apple is apocryphal, very likely made up by Newton himself in later years.

16. Of course, by the same token Kepler's biography "proves" that great astronomers believe in astrology . . .

17. David Hull, *Darwin and His Critics: the Reception of Darwin's Theory of Evolution by the Scientific Community* (University of Chicago Press, 1983).

18. Incidentally, most people think of deduction as what Sir Conan Doyle's fictional character Sherlock Holmes does. In fact, Holmes uses a method of convergence of inferences, precisely what Whewell suggested. Conan Doyle was simply not up to date with the philosophical jargon. See M. Pigliucci, "Elementary, Dear Watson!" *Skeptical Inquirer* 27 (2003): 18–19.

19. In S. Weinberg, *Dreams of a Final Theory* (Pantheon, 1992), 166–90.

20. In D. Dennett, *Darwin's Dangerous Idea* (Simon and Schuster, 1996), 21.

21. See C. P. Snow, *The Two Cultures and the Scientific Revolution* (Oxford University Press, 1959/1998).

22. See the excellent book by J. Kaplan, *The Limits and Lies of Human Genetic Research: Dangers for Social Policy* (Routledge, 2000).

23. See H. Chang, "Complementary Science—History and Philosophy of Science as a Continuation of Science by Other Means," in *Inventing Temperature: Measurement and Scientific Progress* (Oxford University Press, 2004), 235–50.

24. See R. Novella, "Out-of-Body Experiences and the Astral Hypothesis," parts 1 and 2, *New England Journal of Skepticism* 5(2) (2002) and 6(1) (2003), online at http://www.theness.com.

25. See E. O. Wilson, (1998) *Consilience: The Unity of Knowledge* (Knopf, 1998).

Chapter Ten

1. In *Social Studies of Science* 11 (1981): 3–10. Remarkably, and to his credit, Collins more recently softened his critical stance of science significantly, publishing a short essay entitled "We Cannot Live by Scepticism Alone," *Nature* 458 (2009): 30–31.

2. The use of this terminology follows the excellent book by Ronald N. Giere: *Scientific Perspectivism* (University of Chicago Press, 2006). The fact that postmodernism and deconstructionism end up being difficult to distinguish from constructivism simply speaks to the terminological and, more importantly, conceptual confusion that mars discussion in this area. A severe critique of constructivist methodology in general, with several examples from authors that I cite in this and the next chapter, can be found in: Nicholas Shackel, "The Vacuity of Postmodernist Methodology," *Metaphilosophy* 36(3): 295–320.

3. By this word I most certainly do not mean the sociopolitical ideas of Ayn Rand, as expressed in volumes such as *Atlas Shrugged*.

4. See Stephen Hawking, *A Brief History of Time: From the Big Bang to Black Holes* (Bantam, 1988).

5. In Richard Dawkins, *The God Delusion* (Mariner, 2006).

6. A skeptical but comprehensive treatment of scientism can be found in: T. Sorell, *Scientism: Philosophy and the Infatuation with Science* (Routledge, 1991). See also the very balanced S. Haack, "Defending Science—Within Reason," *Principia* 3 (1999): 187–211.

7. For a rather technical, but highly enlightening discussion of the epistemological wars in philosophy, humanities, and science, see S. Blackburn, *Truth: A Guide* (Oxford University Press, 2005).

8. Quoted in R. Youngson, *Scientific Blunders: A Brief History of How Wrong Scientists Can Sometimes Be* (Carroll & Graf, 1998), 13.

9. Ibid., 195.

10. Ibid., 31.

11. Details in Youngson's *Scientific Blunders*, chapter 1, "Cosmology and Earth Science."

12. This, incidentally, started a whole niche genre of physicists attempting to deny the theory of evolution, one of the most recent examples being the cosmologist Fred Hoyle, who was, like his esteemed predecessors, flatly wrong. For more on why Hoyle in particular was wrong, see M. Pigliucci, "Impossible Evolution? Another Physicist Challenges Darwin," *Skeptic* 8, no. 4 (2001): 54–58.

13. George Darwin still got it partly wrong, since he assumed that the sun was fueled by the newly discovered radon, not by hydrogen. However, he was much closer than Thomson to the truth (with a small *t*), since the latter had hypothesized a star made of coal.

14. A complete and engaging version of the story can be found in the 1955 classic *The Piltdown Forgery*, by J. S. Weiner (one of the scientists who eventually uncovered the truth), later reissued by Oxford University Press with a new introduction and afterword.

15. The article appeared in *Popular Science Monthly* and is quoted in E. Caudill, *Darwinian Myths* (University of Tennessee Press, 1997), 97. Caudill's treatment of eugenics is one of the clearest and most concise available.

16. Obviously, Galton did not know of Mendel's work when he began writing about eugenics, but twentieth-century eugenicists saw themselves as applying the principles of genetics to human welfare. This does not mean that there weren't prominent geneticists who criticized the eugenic movement, for instance Thomas Hunt Morgan, one of the most influential biologists of the first part of the twentieth century, famous for his work on the breeding of fruit flies.

17. Caudill, *Darwinian Myths*, 96.

18. Ibid., 104.

19. Ibid.

20. See Paul Lombardo, "Eugenic Sterilization Laws" available at http://www.eugenicsarchive.org/html/eugenics/essay8text.html.

21. Caudill, *Darwinian Myths*, 102.

22. B. Forrest and P. R. Gross, *Creationism's Trojan Horse: The Wedge of Intelligent Design* (Oxford University Press, 2004).

23. All citations from Caudill, *Darwinian Myths*, 106–7.

24. See chapter 1 of M. Pigliucci, *Denying Evolution: Creationism, Scientism and the Nature of Science* (Sinauer, 2002).

25. Caudill, *Darwinian Myths*, 112.

26. Lee Silver, *Remaking Eden* (Weidenfeld & Nicholson, 1998), 4.

27. Ibid., 9.

Chapter Eleven

1. A. D. Sokal, "Toward a Transformative Hermeneutics of Quantum Gravity," *Social Text* 14 (1996): 217–52.

2. In N. Koertge, ed., *A House Built on Sand: Exposing Postmodernist Myths about Science* (Oxford University Press, 1998), 11.

3. Sokal, incidentally, wrote a more recent follow-up to his role in and views of the science wars: *Beyond the Hoax: Science, Philosophy and Culture* (Oxford University Press, 2008).

4. *Social Studies of Science* 11 (1981): 3.

5. Latour, like sociologist Harry Collins (see note 1 for chapter 10), recently softened considerably his criticism of science. For instance, in "Why Has Critique

Run out of Steam? From Matters of Fact to Matters of Concern" (*Critical Inquiry* 30 [Winter 2004]), he wrote: "Entire Ph.D. programs are still running to make sure that good American kids are learning the hard way that facts are made up, that there is no such thing as natural, unmediated, unbiased access to truth, that we are always prisoners of language, that we always speak from a particular standpoint, and so on, while dangerous extremists are using the very same argument of social construction to destroy hard-won evidence that could save our lives." Wow.

6. Koertge, *A House Built on Sand*, 11.

7. For an in-depth criticism of Latour's misunderstanding of relativity, see John Huth's "Latour's Relativity," chapter 11 in Koertge, *A House Built on Sand*.

8. Koertge, *A House Built on Sand*, 101.

9. Ibid., 106.

10. Sandra Harding, "Value-Free Research Is a Delusion," *New York Times*, 22 October 1989.

11. Barry Barnes and David Bloor, "Relativism, Rationalism and the Sociology of Knowledge," in *Rationality and Relativism*, ed. Martin Hollis and Steven Lukes (Basil Blackwell, 1982).

12. In order to get Feyerabend's own version of his story, the reader may wish to read his autobiography, interestingly entitled *Killing Time* (University of Chicago Press, 1996).

13. Paul Feyerabend, *Against Method* (Verso, 1979), 141.

14. See A. F. Chalmers, *What Is This Thing Called Science?* (Open University Press, 1999), chapters 10 and 11.

15. The essay has been reprinted in *The Philosophy of Expertise*, edited by E. Selinger and R. P. Crease (Columbia University Press, 2006). All quotations are from this version of the essay.

16. As I noted earlier, it is interesting that Copernicus's calculations initially didn't get better results than those based on the Ptolemaic system. This was because Copernicus assumed that the planets orbited in circles, since the circle was the traditional "perfect" geometric figure, and it was inconceivable that celestial bodies would move in anything but a perfect way. It took Kepler's insight and metaphysical courage to seriously consider the alternative we now know to be true: planets orbit along ellipses, one of the foci of which is occupied by the sun (which is therefore not exactly at the center of the solar system).

17. T. S. Kuhn, *The Structure of Scientific Revolutions*, 2nd ed. (University of Chicago Press, 1970).

18. The photoelectric effect is a phenomenon by which matter emits electrons after having absorbed energy. It was originally discovered by Becquerel in 1839, but it was Einstein who, in 1905, demonstrated how the photoelectric effect implied that light comes in discrete units called quanta. In 1922 he was awarded the Nobel Prize for this work (not, as many people believe, for his theories of special and general relativity).

19. Helen Longino, *Science as Social Knowledge* (Princeton University Press, 1990). See also her more recent *The Fate of Knowledge* (Princeton University Press, 2001).

20. Longino, *Science as Social Knowledge*, 221–22.

21. See R. Giere, *Scientific Perspectivism* (University of Chicago Press, 2006).

22. The example about to be presented is used in a nice explanation of Bayes's rule at http://library.thinkquest.org/10030/4pceapbr.htm?tqskip1=1&tqtime=0106.

Chapter Twelve

1. This is in part because Socrates and colleagues in the Charmides debate something that the ancient Greeks called "sophrosyne," a concept for which there is no direct equivalent in modern English. Roughly speaking, "sophrosyne" refers to the ability to know and accept one's natural limits, as embodied in the two Delphic mottos "Know thyself" and "Nothing in excess." See Plato, *The Collected Dialogues*, ed. E. Hamilton and H. Cairns (Princeton University Press, 1985), 99.

2. Hamilton and Cairns, *Collected Dialogues: Charmides*, 170e–171c, 117.

3. K. Anders Ericsson and Jacqui Smith, "Prospects and Limits of the Empirical Study of Expertise: An Introduction," chapter 1 in *Toward a General Theory of Expertise: Prospects and Limits* (Cambridge University Press 1991).

4. I am, of course, most certainly not invoking any sort of mystical explanation for expertise. Social scientists have for some time regarded "intuition" as a result of subconscious processing of information by the brain, not as an inspiration from the gods. For a brief commentary on this, see S. J. Segalowitz, "Knowing Before We Know: Conscious versus Preconscious Top-Down Processing and a Neuroscience of Intuition," *Brain and Cognition* 65, no. 2 (November 2007): 143–44.

5. Barbara J. Daley, "Novice to Expert: An Exploration of How Professionals Learn," *Adult Education Quarterly* 49, no. 4 (1999): 133–47.

6. Ibid., 138.

7. Ibid., 142.

8. G. Leinhardt, "Math Lessons: A Contrast of Novice and Expert Competence," *Journal for Research in Mathematics Education* 20, no. 1 (1989): 52–75.

9. Ibid., 74.

10. J. Shanteau, "How Much Information Does an Expert Use? Is It Relevant?" *Acta Psychologica* 81 (1992): 75–86.

11. P. E. Ross, "The Expert Mind," *Scientific American*, 24 July 2006.

12. This is apparently an urban myth that is hard to eradicate. I am still told by colleagues that studying math is good because it improves one's logical abilities. Studying math certainly has benefits, but apparently improved logical thinking (outside of math) isn't one of them.

13. L. White and A. Taket, "The Death of the Expert," *Journal of the Operational Research Society* 45, no. 7 (1994): 733–48.

14. Ibid., 733.

15. In Adams's novels the hero adopts a peculiar "method" to solve crimes: just wait and see. Since everything is connected to everything else, eventually the right clue will cross your path, and you will solve the mystery. Highly recommended as entertaining reading, not so much as a serious philosophy of knowledge.

16. White and Taket, "Death of the Expert," 737.

17. Ibid., 736.

18. Ibid., 735.

19. Ibid., 741.

20. D. Hume, "Of the Will and Direct Passions," part 3 in *A Treatise on Human Nature*. Available in full text at Project Gutenberg: http://www.gutenberg.org/etext/4705.

21. See A. Damasio, *The Feeling of What Happens: Body and Emotion in the Making of Consciousness* (Harcourt Brace & Co., 1999).

22. White and Taket, "Death of the Expert," 735.

23. Ibid., 736.

24. Ibid., 741.

25. See A. I. Goldman, "Experts: Which Ones Should You Trust?" in *The Philosophy of Expertise*, ed. E. Selinger and R. P. Crease (Columbia University Press, 2006).

26. Ibid., 18.

27. Ibid., 21.

28. Socrates had a particular disdain for the Sophists, philosophers who—for a fee—went about teaching people how to effectively defend any argument, regardless of its soundness. While Sophism was, in reality, a significant philosophical school in its own right, modern lawyers may trace their trade all the way to the target of Socrates' criticisms, and we all know the many jokes in circulation based on the shadiness of the legal profession . . .

29. See M. Pigliucci, "The Twilight Zone of Antiscience," *Free Inquiry* 28, no. 3 (2008): 57–58. The book is: *Darwin Day in America: How Our Politics and Culture Have Been Dehumanized in the Name of Science*, by J. G. West (Intercollegiate Studies Institute, 2007).

30. Goldman, "Experts," 23.

31. From Sir Arthur Conan Doyle, *The Memoirs of Sherlock Holmes* (1893).

32. From Sir Arthur Conan Doyle, *The Adventures of Sherlock Holmes* (1892).

33. On Conan Doyle's strange mix of science and pseudoscience, see by Massimo Polidoro, *Final Seance: The Strange Friendship between Houdini and Conan Doyle* (Prometheus Books, 2001).

34. Specifically, as we have seen: K. R. Miller, *Finding Darwin's God: A Scientist's Search for Common Ground between God and Religion* (Harper Collins, 1999);

and M. J. Behe, *Darwin's Black Box: The Biochemical Challenge to Evolution* (Free Press, 1996).

35. Difficult, but not impossible: perhaps the best known agnostic who was a critic of evolution (but not a supporter of creationism) was the cosmologist Fred Hoyle. For a detailed critique of his flawed ideas about evolution, see M. Pigliucci, "Impossible Evolution? Another Physicist Challenges Darwin," *Skeptic* 8 (2001): 54–58.

36. For example, in N. Shanks and K. H. Joplin, "Redundant Complexity: A Critical Analysis of Intelligent Design in Biochemistry," *Philosophy of Science* 66 (1999): 268–82.

37. This is a classic counter to much mystical nonsense. For instance, if psychics can tell the future, why did a major psychic network not predict its own bankruptcy? See http://csicop.org/articles/april_fools_1998/.

Conclusion

1. The interested reader can find a good and approachable summary in an article by Sven Ove Hansson at http://plato.stanford.edu/entries/pseudo-science.

Index